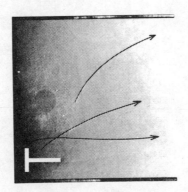

THE UNIVERSITY OF WESTERN ONTARIO
SERIES IN PHILOSOPHY OF SCIENCE

VOLUME 11

BASIC PROBLEMS IN METHODOLOGY AND LINGUISTICS

PART THREE OF THE PROCEEDINGS
OF THE FIFTH INTERNATIONAL CONGRESS OF
LOGIC, METHODOLOGY AND PHILOSOPHY OF SCIENCE,
LONDON, ONTARIO, CANADA–1975

Edited by

ROBERT E. BUTTS

The University of Western Ontario

and

JAAKKO HINTIKKA

The Academy of Finland and Stanford University

D. REIDEL PUBLISHING COMPANY
DORDRECHT-HOLLAND/BOSTON-U.S.A.

Library of Congress Cataloging in Publication Data

International Congress of Logic, Methodology, and Philosophy of Science, 5th, University of Western Ontario, 1975.
　　Basic problems in methodology and linguistics.

　　(Proceedings of the Fifth International Congress of Logic, Methodology, and Philosophy of Science, London, Ontario, Canada, 1975; pt. 3) (University of Western Ontario series in philosophy of science; v. 11)
　　Bibliography: p.
　　Includes index.
　　1. Science—Methodology—Congresses. 2. Science—Philosophy—Congresses. 3. Linguistics—Congresses. I. Butts, Robert E. II. Hintikka, Kaarlo Jaakko Juhani, 1929–　III. Title. IV. Series: University of Western Ontario. The University of Western Ontario series in philosophy of science; v. 11.
Q174.I58　1975a pt. 3　　[Q174]　501s　[502′.8]　　77-22432
ISBN 90-277-0829-0

The set of four volumes (cloth) ISBN 90 277 0706 5

Published by D. Reidel Publishing Company,
P.O. Box 17, Dordrecht, Holland

Sold and distributed in the U.S.A., Canada, and Mexico
by D. Reidel Publishing Company, Inc.,
Lincoln Building, 160 Old Derby Street, Hingham,
Mass. 02043, U.S.A.

TABLE OF CONTENTS

PREFACE   vii

OFFICERS OF THE DIVISION   ix

PROGRAMME COMMITTEE   ix

CHAIRMEN OF SECTIONAL COMMITTEES   ix

LOCAL ORGANIZING COMMITTEE   x

I / PROBLEMS IN THE METHODOLOGY OF SCIENCE

C. A. HOOKER / Methodology and Systematic Philosophy   3
M. V. POPOVICH / Identity by Sense in Empirical Sciences   25
R. WÓJCICKI / Towards a General Semantics of Empirical Theories   31

II / IDENTIFIABILITY PROBLEMS

H. A. SIMON / Identifiability and the Status of Theoretical Terms   43
V. N. SADOVSKY and V. A. SMIRNOV / Definability and Identifiability: Certain Problems and Hypotheses   63
M. PRZEŁĘCKI / On Identifiability in Extended Domains   81
V. RANTALA / Prediction and Identifiability   91

III / FOUNDATIONS OF PROBABILITY AND INDUCTION

T. L. FINE / An Argument for Comparative Probability   105
I. NIINILUOTO / On the Truthlikeness of Generalizations   121
W. C. SALMON / A Third Dogma of Empiricism   149
A. TVERSKY and D. KAHNEMAN / Causal Thinking in Judgment under Uncertainty   167

IV / THE CONCEPT OF RANDOMNESS

C. P. SCHNORR / A Survey of the Theory of Random Sequences   193
R. C. JEFFREY / Mises Redux   213

V / FOUNDATIONAL PROBLEMS IN LINGUISTICS

A. KASHER / Foundations of Philosophical Pragmatics   225
D. WUNDERLICH / On Problems of Speech Act Theory   243

VI / THE PROSPECTS OF TRANSFORMATIONAL GRAMMAR

J. W. BRESNAN / Transformations and Categories in Syntax   261
S. PETERS / Consequence of Speaking   283
J.-R. VERGNAUD / Formal Properties of Phonological Rules   299

INDEX OF NAMES   319

PREFACE

The Fifth International Congress of Logic, Methodology and Philosophy of Science was held at the University of Western Ontario, London, Canada, 27 August to 2 September 1975. The Congress was held under the auspices of the International Union of History and Philosophy of Science, Division of Logic, Methodology and Philosophy of Science, and was sponsored by the National Research Council of Canada and the University of Western Ontario. As those associated closely with the work of the Division over the years know well, the work undertaken by its members varies greatly and spans a number of fields not always obviously related. In addition, the volume of work done by first rate scholars and scientists in the various fields of the Division has risen enormously. For these and related reasons it seemed to the editors chosen by the Divisional officers that the usual format of publishing the proceedings of the Congress be abandoned in favour of a somewhat more flexible, and hopefully acceptable, method of presentation.

Accordingly, the work of the invited participants to the Congress has been divided into four volumes appearing in the University of Western Ontario Series in Philosophy of Science. The volumes are entitled, *Logic, Foundations of Mathematics and Computability Theory, Foundational Problems in the Special Sciences, Basic Problems in Methodology and Linguistics,* and *Historical and Philosophical Dimensions of Logic, Methodology and Philosophy of Science.* By means of minor rearrangement of papers in and out of the sections in which they were originally presented the editors hope to have achieved four relatively self-contained volumes.

The papers in this volume consist of those submitted for publication by invited participants whose work contributes to the general methodology of science, foundations of probability and induction, and foundations of linguistics. Contributed papers in these fields appeared in the volume of photo-offset preprints distributed at the Congress. The full programme of the Congress appears in *Historical and Philosophical Dimensions of Logic, Methodology and Philosophy of Science.*

The work of the members of the Division was richly supported by the National Research Council of Canada and the University of Western Ontario. We here thank these two important Canadian institutions. We also thank the Secretary of State Department of the Government of Canada, Canadian Pacific Air, the Bank of Montreal, the *London Free Press*, and I.B.M. Canada for their generous support. Appended to this preface is a list of officers and those responsible for planning the programme and organizing the Congress.

THE EDITORS

FEBRUARY 1977

OFFICERS OF THE DIVISION

A. J. Mostowski	(Poland)	President
Jaakko Hintikka	(Finland)	Vice President
Sir A. J. Aycr	(U.K.)	Vice President
N. Rescher	(U.S.A.)	Secretary
J. F. Staal	(U.S.A.)	Treasurer
S. Körner	(U.K.)	Past President

PROGRAMME COMMITTEE

Jaakko Hintikka (Finland), Chairman
R. E. Butts (Canada)
Brian Ellis (Australia)
Solomon Feferman (U.S.A.)
Adolf Grünbaum (U.S.A.)
M. V. Popovich (U.S.S.R.)
Michael Rabin (Israel)

Evandro Agazzi (Italy)
Bruno de Finetti (Italy)
Wilhelm Essler (B.R.D.)
Dagfinn Føllesdal (Norway)
Rom Harré (U.K.)
Marian Przełecki (Poland)
Dana Scott (U.K.)

CHAIRMEN OF SECTIONAL COMMITTEES

Y. L. Ershov (U.S.S.R.)	Section I:	Mathematical Logic
Donald A. Martin (U.S.A.)	Section II:	Foundations of Mathematical Theories
Helena Rasiowa (Poland)	Section III:	Computability Theory
Dagfinn Føllesdal (Norway)	Section IV:	Philosophy of Logic and Mathematics
Marian Przełecki (Poland)	Section V:	General Methodology of Science
J.-E. Fenstad (Norway)	Section VI:	Foundations of Probability and Induction
C. A. Hooker (Canada)	Section VII:	Foundations of Physical Sciences

Lars Walløe (Norway)	Section VIII:	Foundations of Biology
Brian Farrell (U.K.)	Section IX:	Foundations of Psychology
J. J. Leach (Canada)	Section X:	Foundations of Social Sciences
Barbara Hall Partee (U.S.A.)	Section XI:	Foundations of Linguistics
R. E. Butts (Canada)	Section XII:	History of Logic, Methodology and Philosophy of Science

LOCAL ORGANIZING COMMITTEE

R. E. Butts (Philosophy, the University of Western Ontario), Chairman

For the University of Western Ontario:

Maxine Abrams (Administrative Assistant)

R. N. Shervill (Executive Assistant to the President)

G. S. Rose (Assistant Dean of Arts)

R. W. Binkley (Philosophy)

J. J. Leach (Philosophy)

C. A. Hooker (Philosophy)

J. M. Nicholas (Philosophy)

G. A. Pearce (Philosophy)

W. R. Wightman (Geography)

J. D. Talman (Applied Mathematics)

J. M. McArthur (Conference Co-ordinator)

For the City of London:

Betty Hales (Conference Co-ordinator)

For the National Research Council of Canada:

R. Dolan (Executive Secretary)

I

PROBLEMS IN THE METHODOLOGY OF SCIENCE

C. A. HOOKER

METHODOLOGY AND SYSTEMATIC PHILOSOPHY

1. INTRODUCTION

Method describes a sequence of actions which constitute the most efficient strategy to achieve a given goal; *Methodology* describes the theory of such sequences.

This statement above is neither a usual formulation, nor agreed to, in the extant literature, but I am going to argue both that it is nonetheless an accurate description of otherwise conflicting approaches to methodology and that it opens the way for a new and much more powerful approach to methodology.

According to the dominant english-speaking philosophic and scientific tradition method is determined by theory of science and more particularly by epistemology. According to another tradition, method defines the scope of epistemology and determines the form of the theory of science. With the former view we associate the empiricist mainstream of this century, with the latter the pragmatists (e.g. Dewey) and Popper. Both approaches can accept the dictum that it is method which distinguishes the practice of science from, say, myth.

I shall later defend a modified version of the science-as-method view by arguing that this is the only acceptable approach for a coherent, systematic realism. But my account of realism will also make it clear how far the conception of systematic methodology required by an adequate realism must transcend the extant conceptions of method and what a more adequate, if more complex, relation to epistemology would consist in.

2. RECAPITULATION: METHOD IN AN EMPIRICIST FRAMEWORK

Elsewhere (Hooker, 1975a) I have presented a more detailed construction of the philosophy and meta-philosophy of empiricism (and of positivism as a more restricted variant – namely, with logic restricted to

Butts and Hintikka (eds.), Basic Problems in Methodology and Linguistics, 3–23.

finite truth functions only). Rather than re-present the abstracted, axiomatized system, I shall summarize the pertinent features as follows: Empiricists aim at the maximal accumulation of maximally justified true beliefs; these latter are to be expressed in a logically clarified language comprising logical terms and whose only meaningful descriptive terms are those which are analyzable into immediately experienceable terms and logical terms. The logical machinery here is not restricted in complexity, though historically the first order predicate calculus has been the dominant system. Call this language $\mathcal{L}$. This language grounds the well-known empiricist epistemology. There is a supporting meta-philosophy which requires a foundationalist epistemology and the statement of philosophical doctrine in terms of the logical structure of a logically clarified language together with the claim that epistemology, theory of language and rationality are independent of, prior to and normative for the remaining structure of human knowledge. Taken together with other elements of a complete doctrine it leads to the following theory of science. *Aim*: The aim of science is the maximization of empirical knowledge, i.e. of maximally evidentially supported true belief. *Demarcation*: The sentences of empirical science are all and only those of the language $\mathcal{L}$. *Scientific language*: $\mathcal{L}$. *Scientific theory*: A deductively axiomatized class of sentences in $\mathcal{L}$. *Scientific Data*: The class of basic sentences in $\mathcal{L}$. *Criteria of Scientific Acceptability*: Minimally adequate evidential support (as specified in the generalized logical theory of inference). *Methodology*: Maximize scientific data, accept all theories that are at least minimally adequately evidentially supported by the data. *History of Science*: Continuous expansion of scientific data, acceptance of progressively more general and successively inclusive theories.

The meta-philosophy and philosophy of empiricism commits it to its conception of methodology, as sketched in the theory of science. Empiricist epistemology presents an absolute foundation for knowledge and the meta-philosophy demands the construction (and reconstruction) of science on its terms; precisely how this reconstruction is to proceed is given on the one hand by the aim of science, itself fixed by the theory of rationality, its nature in turn constrained by the metaphilosophy, and on the other hand by the theory of language which specifies the class of admissible constructions. The details are less important than the overall form, every empiricist position asserting a foundational epistemology with a supporting semantics and meta-philosophy is committed to reconstructing methodology in this form.

3. The Role of Method in an evolutionary Approach: Naturalistic Realism

Once we 'saw' dimly through the 'eyes' of a protozoan, as Campbell (1973) says, and no revelation has been granted us since. Evolution since that time has been total, an evolution of capacity to see, of the capacity for symbolic processes, thought, language. However we came by our philosophy, our science, our culture, we came by it all in the same manner. Humankind emerged from the evolutionary background and was capable, eventually, of all of these things; there has been no revelations, inner or outer. The brain of *Homosapiens* may have been designed through natural selection to fit with the world in some ways, more or less imperfectly, but no one told us that authoritatively, neither an external deity nor an inner voice of reason. Language emerged only gradually, a bewilderingly complex abstraction from the processes in the nervous system, which produces a usefully simple result for social action in the commonsense world, we understand its nature and status even now only very imperfectly; there are no givens here, no revelation. We came by everything then in the same manner; which manner? By trial-and-error guesswork, by throwing out a framework to grasp the world and modifying it as we went along, by blindly experimenting with the creation of societies, language, cultures, driven by ecological pressures and inner needs, learning only painfully slowly what 'works'. And this too is how we came by philosophy – by epistemology, ethics, metaphysics and the rest – through blind guesswork and selective elimination, by something like the same process of natural selection which forms the basis of biological evolution itself.

In such circumstances method becomes central to defining epistemology, for method is all that we have. If one knows in advance the nature of true knowledge (e.g. that it is grounded in sensory givens, or in mystical access to Platonic heavens) then an epistemology may be erected in these terms (using this knowledge!) and method deduced as

the most efficient means to maximise knowledge thus understood. But
if one knows nothing in advance, not even what it is to know, then the
only thing that can matter is a study of the methods of relieving
ignorance. Of these latter there is only one in a universe which is not
given to its organisms once-for-all, the method of conjecture and
refutation.

And this method is 'natural' to our universe, for we have increasing
reason to believe that it is but a representation, at the abstract-symbol
using level, of a basic biological development process that extends
throughout the spectrum of living organisms, from the amoeba to
humankind; that it is the fundamental problem-solving mode of evolv-
ing life. Campbell (1973) has argued this, following Popper (1973) and
a host of others (recently see e.g. Lorentz, 1976; Geist, 1976; Wilson,
1975).

Now epistemology becomes defined in terms of this method. How is
it possible to know? Answer this: How is it possible to pursue such a
method? What is the scope of knowledge? That depends on the
potential scope of the method. When do we know? What does the
method tell us about the conditions under which it works? What do we
know best? What does the method say about unrevisable results, the
relative status of theory and observation, etc.?

Epistemology is defined in terms of the method, but not reduced to
it. There is a great deal else that one must know before answers to
these questions can be given. What sort of creatures are we? What
capabilities have we? What is the nature of societies and cultures?
What role do they play in the development of science? Until we know
a great deal about our perceptual systems, brain functioning, languages, social interactions and so on we cannot say much about the
nature, scope and conditions of human knowing. All of these things lie
outside of method, they belong first to science proper and then to a
fully developed epistemology as a theory of knowing for creatures such
as we are.

There is much science that must be known before epistemology can
be adequately developed, many results of method must be in hand
before the nature of the method itself is understood. Is the form of a
circle apparently detectable here objectionable? Not at all. Historically
we are faced with the simultaneous emergence of both science and

epistemology, conjectural theses in dynamic interaction with each other, more a pair of inter-locked spirals than a circle. That is how it is for those who evolve from ignorance.

But many more questions ought to be asked concerning such a conception of the role of method. An adequate preparation properly requires an explicit statement of philosophy and metaphilosophy of an Evolutionary Naturalistic Realism (ENR). This has been done elsewhere (Hooker, 1975). Briefly, ENR denies the existence of any 'First Philosophy' and claims that the purpose of philosophy of science is to provide a critical theory of theorising with the same epistemic status as any scientific theory. It is in its critical role that its normative status lies. The rational person so acts as to maximise her potentials and the rational society acts so as to maximise its collective potentials. Science is a principle means for the achievement of both types of ends. Science is thus to be viewed as a socially realised (institutionalized) collective method for the enhancement of the potentials of the species. It is one of our survival strategies, to be integrated with both ethology/evolutionary biology and brain function.

From the species standpoint, it is the *collective, historical* human experience which counts. This is so on two counts: (i) The development of theory in these respects centrally involves the development of suitable cultures that foster the development of human beings, individually and collectively. One very important (if thoroughly partial) aspect of this is the notion of a critical culture (cf. chiefly Popper's exposition), and critical not just with respect to factual belief, but with respect to values (ends), means, conceptual schemes-everything. The theory of human ends is at least a theory of human cultures. But cultures are collective creations. (ii) Moreover, the adequate testing of cultural theories may require millenia rather than years. (Are we really historically in a position to decide between a Zen Buddhist or Tibetan culture and our own? – even with respect to exploration of the real nature of the universe?) Certainly it will typically require temporal spans much longer than an individual lifetime. The more so, since there is scarcely an extant (or past) culture that has tolerated critical enquiry to any significant degree yet (one of the reasons why we have never attempted a 'meta-cultural' rapport with Zen Buddhist or Tibetan culture in the spirit of critical, comparative exploration). We are only just beginning, as a species, on the exploration of culture, and hence value theory – and beginning blindly at that. How this view integrates with the general theory of method will shortly become even more obvious.

4. METHOD AND JUSTIFICATION: THE DIVERSITY OF SCIENTIFIC AIMS

There is an intimate connection between the theories of methodology and justification in science, whatever the view of science. The general strategy of the connection is this: if method is the most efficient route to the goal of science and a scientific theory is justified or not on the basis of its achieving that goal, then the core of any justification will be its appeal to method correctly carried out (with the theory in question as result). However, the particular connections established in the cases of empiricism and NR are radically different in nature.

Within an empiricist framework, there is apparently only one goal of science, empirical truth. When this is cashed out in detail, it comes to acquisition of evidentially supported empirical theories. Methodology is a theory about the obtaining of maximally informative and unbiased empiricist data from various situations, together with the appropriate inductive theory. A theory is justified if the relevant data have been acquired in accord with the method and it turns out after proper application of the inductive logic that the theory is sufficiently evidentially supported (whatever this amounts to). In the case of positivism the appeal to method is at once very clear – it concerns only the acquisition of basic data (nothing is left above this but the formation of truth functions) – and very obscure because it requires the introduction of some theory of the process of acquisition of basic statements, about which positivists have had little or nothing to say. (It requires introducing either a psychology of observation, which is anathema to their meta-philosophy, or a theory of conventional choice, which must remain a mystery on their meta-philosophy.) Empiricist methodology is faced with the same problem, to give an account of the foundations of their methodology which is itself rationally justified and consistent with their meta-philosophy. (Cf. Hooker, 1976 and § 8 on meta-philosophical questions in Hooker, 1975a.) In addition, empiricist methodology must include a theory of inductive support and criteria of sufficient inductive support to warrant rational acceptance as knowledge – these too are controversial and hitherto uncompleted tasks.

For ENR methodology is not dictated by an already fixed epistemological goal, rather method follows from a specification of the life-goals we have for science: individual, national and for the race; it serves as part of a theory of epistemology. By contrast with empiricism (and most other philosophies of science, e.g. Popper's), ENR (i) counts the goals for science as strictly on a par with all other life-goals, indeed as inextricably bound up with them and therefore (ii) explicitly recognizes a multiplicity of genuine goals for the scientific activity. Both differences are worth elaborating upon.

On (i). Rational individuals and communities, and the race as a whole, can be regarded as possessing systems of goals which entail 'life strategies'. (Or at least they behave in such a manner as is economically explainable in this fashion; clearly the notion of rational choice which is at the back of these conceptions becomes increasingly subtle and indirect as one proceeds to numerically larger communities and longer time spans and its connections with biological evolution become increasingly important – cf. Section 3 above.) Amongst such goals will certainly be relief from ignorance as an epistemic goal unto itself, but we will also find such goals as survival (security) – short and long term – technological mastery, self-expression, theoretical fecundity, excitement, sensual satisfaction, linguistic economy, control (power), etc. etc. We can introduce some order into this bewilderingly complex array of goals by pointing out that some of them are derivative upon others (eg. the desire for technological mastery as means to achieve control and control as a means to achieve security) and hope to arrive in the end at some hierarchial ordering of goals (as e.g. Fried, 1970).

But this ordering will not be simple, because many of the goals will conflict with each other in a variety of ways; eg. I might seek technological mastery in order to enlarge the means of my self-expression, so long as the latter doesn't conflict with the achievement of a certain level of excitement and the avoidance of unchallenging labour; for the latter I aim at highly economical theories, so long as they are aesthetically pleasing as a means of self-expression and theoretically fecund and so exciting, though these theories must also lead to technological mastery etc. etc. (Just how complex these orderings may be can be ascertained from recent discussion, e.g. Hooker *et al.* (1975).)

The inter-relations among these goals are such, it seems to me, as to preclude separating them out in any non-arbitrary fashion as goals specifically pertaining to the pursuit of science and other goals. Science may be pursued with almost any goal ultimately in mind, from the contemplation of the creator to the lust for criminal wealth; conversely every epistemic goal inevitably leads to the pursuit also of some non-scientific activities (eg. the cultivation of certain social forms as conducive to the activity of critical enquiry). Moreover, within the system of goals it always turns out that there are trade-offs to be made both among epistemic goals and between these and other goals. To attempt nonetheless to separate off a goal or goals for science is surely just to ignore science as a human social endeavour.

The dominant philosophical tradition, however, still regards science as a separate unique endeavour and one, moreover, to be defined quite abstractly. (Positivists, empiricists and Popperians all agree on this.) Science becomes a practical social activity only in an attenuated theoretical sense and a distinctively human activity in no sense at all. (*Would* all possible intelligent creatures in the universe practice science as we do?) The inevitable result is the attempt to segregate the goals of science so as to achieve a characterization of scientific method independent of all other considerations.[1]

The conventional isolationist view of science leads to the formalist approach to the problem of induction (e.g. Lakatos, 1968). I advocate a formulation of the problem in which the system of life-goals and a given data base (together with its scientific history) are fed as input to decision theory and a strategy of acceptance generated as output. The strategy may include more than one theory, include the criticism of the initial data base and concern a group of persons rather than an individual (cf. the discussion in Section 5). The possibility of criticism of the data base shows the need for a feedback loop from output to input, via a detailed theory of science (e.g. a detailed version of

ENR – cf. Hooker, 1974, 1975) so that output can modify input. The loop is activated until a steady state is reached and re-activated each time the data base is changed – cf. Harper (1975). In this fashion the multiplicity of the goals of science leads to a corresponding complexity to the structure of acceptance, for acceptance is always acceptance for certain purposes, relief of ignorance is always relief in certain respects.

If we construe both orthodox empiricist inductive theory and Popperian methodology as offering decision-theoretic accounts of rational choice in science then we can begin to see how to reconcile, or at any rate relate, the two. Both approaches offer goals for science, but these goals turn out to be distinct. Empiricism offers cautious generalization on the evidence, consistency, completeness and the like as criteria, criteria aimed at predictive reliability, technological security and control as goals; Popperianism offers degree of content, falsifiability and the like as criteria aimed at theoretical depth, fecundity and explanatory power as goals. These goals may be incompatible in the sense that in general a given decision could not be optimal for both kinds of goals simultaneously, but there is also nothing to stop a species pursuing them both through different members or the same person pursuing different goals at different times – cf. below. Of course both orthodox empiricism and Popperianism agree on *truth* as the ultimate goal, in some sense, but this turns out to be innocuous; it is not the infinitely long term goal that counts but the more immediate goals to be pursued in the hope of truth that characterizes the differences between the two. If there is direct conflict anywhere, it lies in conflicting claims as to the most efficient strategy to achieve truth (or anyway approximate truth, an important improvement of the Popperian tradition over orthodox empiricism, if the notion is workable, cf. Miller, 1974, 1975); though I should prefer to reconstrue the conflict as one over what should be the dominant utilities of science, for within the domain of *truth* there is surely superficial or 'low level' and deep truth, a species could aim at both. Nor, of course, need there be any other conflict than this over how to account for historical decisions; if now one now the other approach seems to do more justice to a particular historical situation it will be because the scientist's involved had different utility orderings for science and not because they were irrational.

Throughout the foregoing I have used justification in the traditional sense. It is time now to speak of its dual: discovery. The connection between method and justification which the decision-theoretic account builds can be put this way: the theory of justification is just the retrospective methodological theory of acceptance. But the decision-theoretic account also forges an inevitable connection between retrospective methodology and prospective methodology. Consider this claim: For any scientist S and any utility ranking R which S has, if S accepts theories of kind (character, type) T from among extant theories because doing so is utility-maximal for S in the circumstances, then whatever resources S devotes to theory testing should be devoted to T-kind theories and whatever resources S devotes to the invention of theory should be devoted to the invention of T-kind theories. For surely it would be irrational of S to devote resources to the testing or invention of theories which do not maximally advance the goals for science which S holds. Thus we have for ENR, as for empiricism, that retrospective and methodology coincide. (All of this relativized to a particular time, because goals for science may evolve internally to science and they may be re-arranged as circumstances alter, i.e. as a function of the larger cultural setting.)[2] Once again the diversity of utilities leads to a diversity of resource investment in research programmes.

Perhaps the central problem of classical economic theory is that of the allocation of scarce resources among competing claimants. Once the diversity of goals for science and the coincidence of retrospective and prospective methodology are realized, a similar problem arises in the allocation of scientific research resources. An individual, society, species must adopt by some means some allocation of resources among the various possible, generally competing, research strategies. Just as in economic theory the optimal solution is in general not the allocation of all of a scarce resource to one individual, so in scientific research we cannot expect the optimal strategy will be to do only one thing. Thus we can expect that the normal, rational situation in scientific research to be one in which a diversity of scientific researches are being actively pursued simultaneously. This conclusion holds surely even if the range of admissible goals for science is imagined restricted to something approximating the purely epistemic, for an optimal allocation of resources would still require eg. that at one end of the methodological spectrum some scientists work on the incremental elaboration of the most acceptable theory of the day while at the other end others work on radical alternatives to it. Feyerabend (and others) have provided special reasons why this methodological pluralism is reasonable (translation: is necessary for an optimal strategy), these have to do with the explanatory flexibility of explanatorily powerful and general theories (see e.g. Hooker, 1972).[3]

The conception of Method considered here for ENR is just that needed to display the internal complexity and subtlety of scientific choice. Consider the operative levels of scientific practice (cf. Hooker, 1975b): there is the level of applied theory, including theory of instrumentation, where detailed models are constructed for specific situations; next the level of general theory, wherein the general principles of a theory, not situation-specific, are systematized and developed; next the level of proto-theory at which those background theories too fundamental to normally be questioned are explicitly developed (e.g. the linear theory of time, theory of continuous space); next the level of abstract theoretical framework at which abstract mathematical or logical structures are delineated (e.g. Hilbert space theory, theory of relations); finally the level of systematic ontology, wherein the fundamental forms assumed for the world are made explicit (e.g. atomic, plenum or process systematic ontologies). Although one asks the same kind of question at each level, "In the light of all of the available information, do I accept or reject this theory?", each of these levels involves somewhere different decision-making problems.

At the first level of detailed applied theory where everything in a model situation is specified (cf. again Hooker, 1975b), the major question is: "What is the relation between the data as constructed and the predictions of the theory?" Other considerations have to do with the possibility of arriving at the correct predictions for the wrong reasons (e.g. construction of data using a faulty theory, gratuitously cancelling effects) and the availability of alternative, equally empirically adequate, applied theory. As soon as one passes to the more general level of pure theory a host of additional problems present themselves, especially when the results are negative for the specific applied theory: Could we alter the theory of the instruments?, or of the data construction?, or deny the *ceteris paribus* clause?, or reject one of the auxiliary theories?, etc. And if the results are positive, there are still the same kinds of additional questions as appeared at the applied level, together with the question of how the support is to be distributed among the various components of the applied model which incorporated the general theory, with this latter question re-appearing even more urgently when a class of situations have been tested in each of which the general theory is embedded in an applied theory. At the next higher level one is required to test a variety of general theories in which various proto-theories are embedded before any reasonable decisions can be made, each general theory in turn requiring the testing of a sequence of applied theories. And so on up. Thus, as one ascends still higher toward the increasingly abstract and general these considerations multiply and the number of research avenues which must be explored, each requiring the conducting of long sequences of tests before any kind of reasonable decision can be made, increases sharply.

Thus on this view each of these levels of science is open to testing and rational decision making; the obliqueness of the relationship between data and theoretical level, however, increases sharply as one moves through this list of levels with the consequence that the time period required for the elaboration and testing of a theory at each level also increases sharply. The time frame for coming to an adequately supportable decision on some highly detailed applied theory might be 5 minutes and typically of the order of a year or less, the time frame for adequately exploring the appropriateness of, say, the atomic ontology as the fundamental form for physical theory is certainly longer than 2000 years in our particular case (though none of our cultures have been critical) and could be expected to be of this order of magnitude, even in a critical society. (On the latter notion cf. Section 3 above and further below.)

This internal complexity in the structure of time frames for testing various levels of scientific theorising leads to a corresponding complexity in the structure of rational research resource allocation problems and a sharper statement of the conflict between various goals for science. For a society (e.g. a nation or group of nations) there is the problem of how to allocate resources over time, perhaps over generations, to the various levels of research programmes that it considers worthwhile pursuing. And for the species there is the issue of how to allocate resources over time, say millenia, to the exploration of alternative cultures embodying perhaps alternative conceptions of reality, alternative institutionalizations of the activities of science and (therefore) alternative research programmes at all levels. It seems safe to say that *homo sapiens* has hitherto never approached the latter question and never at all seriously approached the former question, except perhaps for the last two decades and then only in embryonic form. We humans have never experienced a truly critical culture (cf. below Section 5). Such long term goals clearly often conflict with the short term goals of technological security and progress, explanatory enlargement and unification of science within existing theoretical perspective. They conflict also with the middle term goals of elaborating theories of particular abstract types (eg. phase space theories), which embody particular systematic ontologies (e.g. the atomic ontology) etc.

It is certainly the case that hitherto the allocation of research resources has not taken future generations into account, nor yet alternative cultures, nor even alternative societies of institutions within a generation (as far as can be avoided), but rather has proceeded by dint of the strictly locally powerful both in time and social space. But this generates a generalized form of a 'Prisoner's Dilemma' game from the point of view of the species, for each institutional sub-culture pursuing its own goals will inevitably prove to be a less efficient basis for the allocation of research resources than the 'long-term co-operative solution' in which we all contribute optimally to a species strategy. A critical culture would be one which possessed a public rationale and corresponding institutional structure for solving this problem in a rationally justifiable way. (There may be many such ways corresponding to many cultures and many conceptions of the world and of the rational human.)[4]

5. SCIENCE METHODOLOGY AND SCIENCE POLICY:
EVOLUTION OF THE SPECIES

One of the important motivations of the decision-theoretic construal of scientific methodology is its permitting the unification of the normative foundations for scientific methodology and science policy. Of course I did not intend it to follow from this statement that there was never any distinction to be made between science and anything else, e.g. politics, or that political decisions should always be those that dominate in the pursuit of science, or that there never would be any rational conflict between the objectives of scientists and those of other segments of society. Just as the methodological policy of "anything goes" does not hold within scientific methodology it does not hold either for the interaction of a scientific institution with the remainder of the social structure.

Indeed, my intention in seeking this common foundation between methodology and policy is deeper-lying than an immediate concern for politics and the like.

A realist, I have argued, is required to give a complete naturalistic account of the species *homo sapiens* in its full evolutionary setting. It is my increasing intuitive conviction that an adequate biological under-standing of our species suggests that many of our crucial capacities are really collective or communal species capacities rather than intrinsi-cally individual capacities. At the present time the question of group or communal selection processes within evolutionary dynamics is thoroughly controversial and at the forefront of research; nonetheless, we have very strong evidence for my point of view in the case of the structural features of the gene pool for *homo sapiens*, since this pool is a species possession rather than an individual possession, for the ability to actually use language and for any number of other abilities which are clearly social in nature (e.g. the ability actually to solve differential equations). But I suppose that our possession of a concept of self and a correlative notion of reality, our conceptions of rationality, epis-temology and the like are all equally strongly tied to being features of the socializing species rather than simply independent intrinsic proper-ties of individual human beings. And in particular I should want to argue for the view (though not here) that the epistemic capacities of *homo sapiens* are functions of the biological capacities of the species *homo sapiens* for specific forms of social interaction. That we are capable of only certain forms of communication (e.g. speech and not telepathy) and that our evolutionary heritage strongly inclines us toward certain (rather primitive) forms of social organization (cities, voting schemes) has a profound influence upon the epistemological enterprises which *homo sapiens* is capable of pursuing. (And, of course, to borrow Kant's point, a profound correlative influence upon the kinds of epistemic enterprise which *homo sapiens* is capable even of conceiving of pursuing.) The same considerations apply to all of the other characteristics of *homo sapiens* which I mentioned earlier.

Moreover, our learning about our capacities in this fashion is itself part of the growth of science. In the simultaneous evolution of scientific theory and theory of science (and theory of society) this is one of the strongest feedback links from scientific theory to theory of science (and society). It is one of those links which I wish, as a realist, to take very seriously. But if one does so one has to admit that a deep account of the nature and evolution of the scientific enterprise must make reference to the social structure of the organisms pursuing that enterprise. A penetrating account of the nature and history of the scientific enterprise on this planet must take into account not only the evolution of scientific theorizing but the evolution of the biological, institutional and cultural context generally within which the scientific enterprise has been conceived and operationalized. Moreover, from now on it will become increasingly important also to take into account the development of selfconscious theory of the design of institutional forms and cultural systems for the future evolution of the scientific enterprise. For it is the distinctive feature of the 20th century that we have become aware selfconsciously for the first time in the planet's history of the designs or structures or systems, biological and sociological, operative in our circumstances and we are for the first time in a position to debate the choice of our own future designs.

One consequence of this point of view is that there is not, and cannot be, what might be called a purely intrinsic characterization of, or demarcation of, the nature of science. What makes an enterprise a scientific enterprise has to do with the structure of the institution which realizes that enterprise and the structure of the culture generally (value systems, modes of communications etc.) within which that institution is imbedded. There are more or less thoroughly critical, or scientific cultures and these have the capacity to pursue scientific enterprise to a greater or lesser extent, but it follows from the above that there will not be, cannot be, a characterization of that enterprise which makes no reference to the policies, procedures, communicational forms and so on which the institutions in question embody. In terms of purely internal characterization, e.g., I can see no way plausibly to make a distinction between the pursuit of science in western culture and the pursuit of a mythological-magical understanding of the world in some other culture; what seems to me to make the crucial difference is the nature of the social processes set in train by each culture for the pursuit of their point of view.[6]

The integrated approach to rational action described here opens the way for a reformulation of the problem of the social responsibilities of scientists. This problem cannot be adequately stated whilst ever an unbridgeable gulf separates methodology and policy. The scientist's rational behaviour qua scientist is separated and unrelated to his rational activity qua citizen. Under this scheme he is forced to choose between society and science and the choice will inevitably be judged irrational by the other party. In general then no rational reconciliation is deemed possible or, if it has somehow to be asserted, it is assumed achieved mysteriously in the depths of the scientist's spirit. The present framework suggests the following reformulation: In the light of the present total situation of the species (socio-cultural, technological, epistemological etc.), is pursuing research strategy R likely to promote or confine the future developmental capacities of the species? The answer depends, obviously, on the evaluation of the full range of human developmental capacities. It is confined too by our ignorance of them. One part of the answer certainly has to do with the direct impact of pursuing R on the growth of knowledge. But another part has to do with the chain of consequences passing through the resulting socio-cultural, especially institutional, structure and technological/technique structures and reflecting back on our future capacities to pursue understanding, mastery and all of our other goals. There is the most intricate interaction between these parts, in fact they cannot be separated off from one another at all. Whatever the imponderables involved, this is a much more adequate form in which to pose the problems than the usual schizophrenic conception of the issue.[7]

That these processes of decision embrace an entire culture is demonstrated e.g. by the very different approaches to the world taken by other cultures (e.g. Tibetan). Nor need alternative approaches be less critical, I think. Nor need it be obvious that western science is the only effective approach to reality, or the most fruitful in the long run. Indeed, in evolutionary perspective *homo sapiens* has evidently just entered upon an experimental exploration of the resources and potential of differing cultures. This is as yet 'blind' exploration by uncommunicating individuals (i.e. individual cultures). The species has yet to face the challenge of entering the period of the critical exploration of fundamentally different conceptions of reality. Not only the result of such an experiment would be in doubt (for the tens of millenia it would take to accumulate relevant experience – these would be the longest time-frame experiments of all) but it is doubtful whether our species possesses the social and intellectual capacities to even attempt it. *A fortiori* the evolutionary result of our present experiment with western science is in doubt (cf. Hooker, 1975a). At no other time in the development of the species has there been a more urgent requirement to understand the complex relations between methodology and policy and culture and to work towards a viable and humane culture.

The University of Western Ontario

NOTES

[1] This is true of most of those offering a decision-theoretic account of scientific method (e.g. Levi, 1967), though not all (cf. Jeffrey, 1965; Rudner, 1966).

The dominant aim of science according to Levi's influential *Gambling with Truth* is relief of epistemic ignorance. Now this notion may perhaps be capable of being given a unique sense in a positivist setting or other settings equally severe in their restrictions on science, but in the context of NR it is multiply ambiguous. We are epistemically ignorant in many ways, ignorant of The True Theory, ignorant of more theoretically fecund theories than those we have, ignorant of more practically fecund theories than those we have, ignorant of technological mastery over nature in the ways we desire, ignorant of the forms of culture and social institutions best suited to the pursuit of enquiry, ignorant of the social consequences of pursuing the foregoing goals in some particular manner, and ignorant of the best long term system of goals, epistemic and other, for human fulfilment. Which of these ignorances is it the business of science to remove? And in which order? Only by arbitrarily concentrating on those ignorances at the head of the list and ignoring those at the tail of the list can one arrive at a characterization of scientific goals independent of the full system of human goals.

² Of course, that there is a prospective methodology is quite compatible with there not being a 'logic of scientific discovery' in the traditional sense, because in the decision-theoretic framework prospective methodology might be more transparently labelled 'logic of future-oriented choice', for it specifies the choices to be made in the investment of research resources (and so ties methodology to policy), it does not provide a recipe for ensuring the desired outcome. (This distinction shows how Feyerabend can be right about the coincidence between retrospective and prospective methodology and wrong about claiming the non-existence of both on the Popperian grounds that the latter doesn't exist; the argument equivocates between 'logic of discovery' and 'methodology of resource allocation' – see Hooker, 1975a, and below.)

³ Feyerabend backs up this argument with an analysis of historical examples in which, he argues, scientists quite reasonably resorted to radically unorthodox methods before succeeding in having their view recognized and accepted (the favourite is Galileo – see Feyerabend 1969, 1970a, b). Feyerabend's own conclusion is the most extreme version of pluralism: "anything goes" (see Feyerabend, 1970b; cf. Hooker, 1972).

To assess the reasonableness of this methodology, it is important to distinguish the following claims:

C_1: There is a fixed, unique methodology for scientific research.
C_2: There is an adequate, unique theory of prospective methodology, and distinguish among these claims.
S_1: There is a particular, finite class of research strategies which will be adequate to cover individual rational action in every historical situation.
S_2: There is a finite class of patterns of research strategies which are adequate to cover species-wide rational action in every historical situation.

First let us be clear that (i) C_2 does not entail C_1, (ii) nor does S_2 entail S_1, for (i) there may be an adequate and unique theory of methodology which nonetheless doesn't specify a fixed procedure and (ii) the pattern of research resource allocation may indicate a historically conditional allocation of species-wide resources to each of a set of strategies spanning the full spectrum of possibilities, from radical guess-work to cautious variation on the known, such that this pattern even entails the negation of S_1.

And these possibilities would be exactly those we would expect from the diversity of goals for science and the decision-theoretic approach to methodology. Although there is a general theory of methodology, it does not specify a fixed procedure, rather procedure is a function generally of operative utilities and (believed) historical circumstances (rational decisions generally are functions of these two factors). Moreover, the situation of the individual is usually quite different from that of the group here; typically the individual has the resources (time, energy, money etc.) to work on only a few difficult problems in a lifetime and only on one or two at a time; the societal group, however, can pursue many different problems simultaneously and pursue research programmes that stretch over generations. Thus the problem of allocation of resources for the individual will be quite different from that of the group; more particularly, it will in general prove rational for the group to pursue all or most avenues of research, including the radical alternatives, no matter how the group decisions are made. (Even if the probability of success for most utilities decreases as the strategy becomes more radical, the pay-off if a success occurs also increases, so that almost any society with sufficient resources will always have some persons pursuing these approaches.)

Now we are in a position to evaluate Feyeragend's radical pluralism. First the judgement "anything goes" is equivocal as between the individual and the group.

Applied to the individual level, it might be translated as the negation of S_1; applied at the group level it might be translated as the negation of S_2. But there is no reason which Feyerabend supplies to reject S_2, so long as the class of patterns of allocations of research resources among strategies contains members which specify wide-ranging and circumstantially flexible allocations. Moreover, he focuses on cases of individual choice. Notice next that the negation of S_1 is compatible with S_2, and even entailed by it under the present approach to method, and moreover that the negation of C_1 is compatible with C_2, indeed $\sim C_1 \cdot C_2$ is just what one would expect under the present approach to method. Thus Feyerabend's claims about the necessity for great flexibility of method are compatible with the existence of a prospective methodology, even with one that entails a unique group-allocation strategy. Finally, observe that which methodological recommendations are relevant for an individual depends strongly on the socio-cultural setting in which he/she is operating. Galileo operating in a monolithic dogmatic culture arrayed against alternative theories, Feynman operating in an oligopolistic market of saleable ideas, some mythical successor operating in a utopian society of complete co-operation – these are all radically different decision-making contexts. Thus in a fully co-operative, collectively rational society, a given individual might indeed not be rational in adopting the infinitely flexible policy Feyerabend recommends, for there is the assuredness that others will do their part, openly and critically, to contribute to the overall research programme; but Galileo faced with his society, or one of us operating in a market of quasi-popular ideas are faced with very different problems. In these cases, as for the political revolutionary and the perfectly competitive businessman, it may well be that 'anything goes'. Still, this would be primarily a comment upon the structure and dynamics of the socio-cultural setting for the practice of science rather than a reduction of either methodological theory or procedure. In sum, despite Feyerabend's examples and criticisms, C_2 and S_2 remain.

Remain, that is, so long as we adopt the ENR decision-theoretic formulation. It is a damaging commentary on the remarkably simplistic conception of method entertained by orthodox empiricists and Popperians alike (and even Feyerabend in his way) that they have aimed to offer unique, fixed rules of method for the practice of science (Feyerabend's fixed rule was: no rules), instead of a flexible, situationally conditional theory of strategy selection.

[4] It is in this perspective, at any rate, that the transition in the quasi-Popperian school from Popper's methodology to Lakatos' methodology of scientific research programmes constitutes a distinctive advantage by explicitly introducing the historical, evolutionary dimension as part of the evaluatory process. And we may understand the significance of the notion of a research programme as that of the systematic exploration at some level of the serious realization of that level at all lower levels; this exploration pursued in the light of the utilities historically current for science (critically examined in the light of experience). Historically degenerating problem shifts are those producing decreasing pay-offs in the coin of the operative utilities and regenerating shifts conversely. This translation at once re-casts the Lakatos theory in a wider perspective allowing for variation of methodological principles from programme to programme (as utilities vary; in the light of experience) and also corrects the Popperian-type failures of even this wider Lakatos approach (i) to distinguish adequately among levels of research programmes, (ii) to distinguish the multiplicity of ends for the pursuit of a given programme, (iii) to distinguish the evolution of social institutions for the pursuit of research from the evolution of theories.

Moreover it is within this perspective of simultaneous pursuit of long and short term goals by a society or species, a situation where support of a 'mixed strategy' is realized as an actual pursuit of all goals by varying fractions of society, that Feyerabend's methodological pluralism finds its natural rationale.

And so long only as we remember that we are dealing with uncritical cultures, Kuhn's societal-specific normal/revolutionary distinction also finds a natural place as a part of the societal long term strategy for the sequential elaboration and testing of a research programme at a fundamental level (say that of general theory or deeper). It is important that it be sequential (this will be because of the uncritical structure of the social institutions involved), otherwise the rationale for Kuhn's order collapses (i.e. collapses where elaboration of distinct versions of a given level may be deliberately pursued simultaneously).

[5] We can hope that much of the language of mathematical ecology and, through it, of mathematical economics will thus become applicable to institutional theory. Through net theory the mathematics of neurophysiology may also usefully apply (as it does to ecology).

[6] Thus I do not believe that the Popperian program for demarcating science succeeds, or could succeed. Indeed, Popper himself recognizes that the principles for the choice of his basic statements lie outside of the characterization of science which he offers. (For more criticism cf. Hooker, 1976).

[7] It was of course Popper in this century who first clearly drew attention to this social foundation for the scientific enterprise despite his resisting the connecting of the theory of that enterprise with the larger theory of the social human species, despite even the later clear emergence of evolutionary themes in his writing (cf. Popper, 1973; Hooker, 1975a). Subsequently, Feyerabend enlarged substantially on these themes – cf. Hooker (1972), where a substantial bibliography may be found. I have developed some of these themes at greater length elsewhere, see Hooker (1974, 1976).

BIBLIOGRAPHY

Campbell, D. T.: 1973, 'Evolutionary Epistemology', in P. A. Schilpp (ed.), *Philosophy of Karl Popper*, Open Court Publishing Co., La Salle, Illinois.

Feyerabend, P. K.: 1969, 'Problems of Empiricism II', in R. Colodny, (ed.), *Pittsburgh Studies in the Philosophy of Science*, Vol. IV, University of Pittsburgh Press, Pittsburgh.

Feyerabend, P. K.: 1970a, 'Classical Empiricism' in R. E. Butts and J. W. Davis (eds.), *The Methodological Heritage of Newton*, University of Toronto Press, Toronto.

Feyerabend, P. K. 1970b, 'Against Method', in M. Radner and S. Winokur (eds.), *Minnesota Studies in the Philosophy of Science*, Vol. IV, University of Minnesota Press, Minneapolis.

Fried, C.: 1970, *An Anatomy of Values*, Harvard University Press, New Haven, Connecticut.

Geist, V.: 1976, *The Emergence of Homosapiens*, Mimeographed, to appear.

Goodman, N.: 1965, *Fact, Fiction and Forecast*, Bobbs-Merrill, New York.

Harper, W.: 1975, 'Rational Belief Change, Popper Functions and Counterfactuals', in Hooker *et al.*

Hooker, C. A.: 1972, 'Critical Notice: Against Method, P. K. Feyerabend', *The Canadian Journal of Philosophy* **1**, 489–509.

Hooker, C. A.: 1974, 'Systematic Realism', *Synthese* **26**, 409–497.

Hooker, C. A.: 1975a, 'Systematic Philosophy and Meta-Philosophy of Science: Empiricism, Popperian and Realism', *Synthese* **26**, 245–299.

Hooker, C. A.: 1975b, 'Global Theories', *Philosophy of Science* **42**, 152–179.

Hooker, C. A.: 1976, 'Towards an Adequate Conception of Scientific Rationality', to appear.

Hooker, C. A., McLennen, E. and Leach, J. J. (eds.): *Foundations and Applications of Decision Theory*, 2 *vols.*, Reidel, Dordrecht, Holland, Forthcoming 1977.

Jeffrey, R. C.: 1965, *The Logic of Decision*, McGraw Hill, New York.

Lakatos, I.: 1968, 'Changes in the Problems of Inductive Logic', in I. Lakatos (ed.), *The Problem of Inductive Logic*, North–Holland, Amsterdam.

Levi, I.: 1967, *Gambling With Truth*, A. A. Knopf, New York.

Lorentz, K.: 1976, *Die Rückseite des Spiegels*, Pieper, Müchen, 1976.

Miller, D.: 1974, 'Popper's Qualitative Theory of Verisimilitude', *British Journal in the Philosophy of Science* **25**, 166–177.

Miller, D.: 1975, 'The Accuracy of Predictions', *Synthese* **30**, 139–148.

Popper, K. R.: 1934, *Logik der Forschung*, Wien. (Translated as *The Logic of Scientific Discovery*, Hutchinson, London, 1962.)

Popper, K. R.: 1973, *Objective Knowledge*, University Press, Oxford.

Rudner, R.: 1966, *Philosophy of Social Science*, Englewood Cliffs, Prentice-Hall, New Jersey.

Wilson, R.: 1975, *Socio-Biology: The New Synthesis*, Harvard University Press, Cambridge, Mass.

M. V. POPOVICH

IDENTITY BY SENSE IN EMPIRICAL SCIENCES

The considerations proposed below contain an attempt to restrict the notion 'a sense', 'identity of sense' and 'difference of sense', respectively, on the basis of notions borrowed from the theory of information. I believe this way is very simple, and I do not need a lot of time to explain the matter. But at the very beginning I would like to make a few more general remarks concerning the semantical analysis of empirical sciences.

The difficulties connected with the formal definition of the notions 'truth', 'proof', 'identity of sense' in empirical sciences are well known. The problem of the description of the correlation of theory language and observation language is connected with them.

Of course, if we do compare the situation in the present-day logic of the empirical sciences and the situation in the twenties, we can find a very significant progress. But the principal problems, such as the logical reconstruction of a principle of observability, are not solved yet. The problem of the observability of theoretical constructions seems more and more difficult. In any case, such is the situation in the real science, although not in the logic of science.

Fifty years ago many eminent scientists hoped that the destruction of old dogmas and a pure scientific approach can be obtained if and only if we shall take account of some principle of observability and shall not use the notions without their operational definition. Today in principle unobservable notions of the old physics have really been eliminated, but their place is occupied by such immediately unobservable notions as a vector of state or the metric of gravitation—the products of a high degree of mathematical and logical abstraction.

I would like to underline that the extreme kinds of reductionism such as those of Ernst Mach or of the 'Wiener Kreis' were never supported by leaders of natural science of the twentieth century. But the crisis of reductionism does not mean the end of the aim of the

Butts and Hintikka (eds.), Basic Problems in Methodology and Linguistics, 25–30.
Copyright © 1977 by D. Reidel Publishing Company, Dordrecht-Holland. All Rights Reserved.

modern science to find operational definitions for different scientific notions on the basis of experiment and observation and to find in experience the last answers to the scientific questions.

The interpretation of the principle of observability proposed below is not, strictly speaking, a logical one. I shall not use the notions and techniques of logic but only certain general schemes which it is possible to hope to define more exactly. So, the present paper does not contain any attempts to solve well known logical problems; rather, it is an attempt to avoid some of them. Justifying myself I can only say that the logical formulae used in the logic of empirical sciences are in fact very often only such general schemes, not formulae in the strict sense of the word. In order to avoid illusions, let us pay attention to the following assumptions.

(1) When semantics is constructed for the logical languages, sometimes it is said that the propositional variables denote the propositions of a natural language which are considered as 'pieces of information', and so on. I would like to consider propositions that is, 'pieces of information', but it is impossible to do so in the framework of logic.

The variable is only an (empty) place where a constant, i.e. a name of an object of a fixed class, can be written. The sentences of a natural language cannot be substituted for propositional variables, nor the adverbs of a natural language for predicate variables. The illusion arises that when instead of 'John is tall' we write 'Tall (John)', we are at once in the field of the logic.

In order to avoid similar illusions and to speak in some cases about the sense of a proposition of a natural language, two kinds of variables will be distinguished below. If certain fragments of the natural language (for example the propositions) can be substituted for certain variables, then the letters of the Greek alphabet will be used. Let us call it a general scheme Λ. In the logical language $\mathscr{L}$ only the objects of some fixed class (for example the truth-values) can be substituted for the variables; we shall use there, if necessary, the Latin alphabet. A propositional constant in Λ denotes the sense of a proposition. There we can speak about 'pieces of information'. The relation between Λ and $\mathscr{L}$ is the relation of (non-formal) translation. Thus, for example, the formula $f \supset t$ (were 'f' and 't' means respectively 'False' and 'True') in $\mathscr{L}$ cannot be interpreted in Λ by a similar substitution of some propositions, such as 'if $2>3$, then $2\times2=4$'. It must be interpreted non-formally, rather as '$2\times2=4$ even if 2 would be greater than 3'.

(2) The expressions 'F-truth' and '$\mathscr{L}$-truth' can be explications of the expressions 'factual truth (on the basis of observation or experience)' and 'theoretical truth' only with the following additional suppositions: (a) the logical language can be broadened by means of the introduction of suitable terms and predicate constants (not of 'concrete predicates' such as 'Tall (x)', but as we do in formal arithmetic) so that each scientific theory can be adequately expressed in it; (b) all the sentences speaking of the laws of nature will be analytically true in this language. Such an assumption is too strong. Hence we shall regard the difference between the law-like propositions as the difference between the senses of the corresponding propositions in Λ.

1. THE SCHEME OF PROOF

Thus we distinguish clearly the scope of non-formal explanations in terms of a natural language, where we can speak of 'pieces of information', from the scope of the strict semantical interpretations of logical expressions. The next step consists in the construction of a general scheme of the proof. It must be as near as possible to our intuition, and corresponding restrictions must translate it into a strict semantical notion of proof in logic.

Well, let $\mathscr{A}$ be a set of propositions which we shall call 'initial information'. Then let there be a question such that each member of the finite set of its answers $\beta_1, \beta_2, ..., \beta_n$ can be consistently joined to $\mathscr{A}$. If β_i can be consistently joined to $\mathscr{A}$, that means that $\mathscr{A}$ doesn't contain actually $\neg\beta_i$. Using '$\vdash$' as a shorthand for 'from... follows...', '$\vee$' as the shortening of 'or' (not as the corresponding strict logical and metalogical expressions!), let us write down this as

$$\mathscr{A}\vdash\beta_1\vee\beta_2\vee\cdots\vee\beta_n.$$

We shall call the β_i hypotheses (about a law–or fact–the difference is not significant there).

Then let be some proposition γ such as that γ can be consistently joined to $\mathcal{A}$ and

$$\mathcal{A}, \gamma \vdash \beta_1 \vee \cdots \vee \beta_m.$$

If $m < n$, then we shall call γ an argument for or against β_i $(1 \leqslant i \leqslant n)$. If $m = 1$, i.e. if all the hypotheses are rejected except one, we have a proof of β_i.

In other words, we do suppose that a proof in empirical sciences as in deductive sciences consists in the formulation of the finite set of hypotheses or possible answers to certain questions and in the rejection of some of them by means of some arguments. The proof is finished if and only if all hypotheses except one are rejected.

The transition from this general scheme to the different types of semantics for in $\mathcal{L}$ (with different additional conditions) can be realised easily. It is possible also to pass to theoretically-probabilistic systems.

It is vital that this general scheme should seem to be near to our intuitions about empirical as well as deductive proof. We can restrict the way the argument γ is obtained. If *all* arguments are obtained from $\mathcal{A}$ according to some rules, then any proved β_i is analytically true. When we distinguish (by sense) propositions of theory from propositions of fact (T-propositions) we can distinguish from each other (1) a F-proof of a F-hypothesis, (2) a F-proof of a T-hypothesis, (3) a F-proof of a T-hypothesis, (4) a T-proof of a T-hypothesis. Thus in the framework of classical mechanics a theoretical proposition about the equivalence of inertial mass and gravitational mass is proved experimentally, whereas in the framework of relativistic mechanics it is proved theoretically (analytically). It is clear that each theory is built so that a maximum number of T-proposition obtains a T-proof.

As follows from this consideration the problem of the degree of confirmation cannot be solved by means of a comparison of isolated T-propositions and F-propositions. If we want to define the degree of confirmation, we must consider some factual argument in connection with all hypotheses.

2. Semantical information

The next step consists in the definition of semantical information. The quantity of semantical information in γ is defined so as to depend on the fact that the joining of γ to $\mathscr{A}$ influences the probability of each of the hypothesis β_i. This γ bears the maximum of information if it allows us to reject all hypothesis except one. In this case we choose between β_i and all other hypothesis regarded as not β_i; probability *a priori* of β_i (and of $\neg\beta_i$) is 1/2. So it coincides with the usual ideas of a probabilistic theory of information.

3. INDISCERNIBILITY BY SENSE

A few concluding remarks concerning the indiscernibility by sense have perhaps some methodological interest.

Two arguments γ_1 and γ_2 influencing the probability of each hypothesis β_i identically bear exactly one and the same information and, consequently, have one and the same sense as far as this set of hypothesis is concerned. Two arguments influencing identically the probability of each possible proposition in some theory have one and the same sense in this theory.

Let the hypotheses be T-propositions and the arguments-F-propositions. In this case the thesis mentioned above is rather trivial. It means that two experiments (or observations) are one and the same experiment (or observation) if and only if they influence the probability of our hypotheses identically.

It demonstrates the necessity of the ·definition of the notion of 'fact' (and, hence, 'different facts' and 'the same fact') on the basis of theoretical considerations. We are looking at facts from a theoretical viewpoint.

But the situation is perhaps more interesting when two theoretical propositions different from the point of view of common sense are identical from the theoretical (and factual) viewpoint.

If two theoretical assertions influence the probability of each hypothesis (both theoretical and factual) in the frames of the theory by the same way, then they have exactly the same sense.

An example from the history of science can help to explain the matter. What would happen if one fine day all the distances in the world suddenly increased very much so consistently that by means of measurement this change in things, distances, and scales would not be possible to reveal? Once Leibniz took part in the discussion on this question but did not, I believe, expound his position exactly. As long as the assertion about a sudden increase in the scale and distances still influences all F-hypotheses in the same way as the assertion about the unalterability of scales, it seems to be rejected as being an unobservable one. A reductionist understanding of the 'observation principle' apparently implies a similar valuation of theoretical hypotheses in general. But it is not so. Such an assertion has the same sense as an assertion about the unalterability of the world. So it is possible to understand the viewpoint of Leibniz as an assertion about indiscernibility of big and small in the endless Universe as a whole.

Modern physical theories yield closely similar examples. The indiscernibility of T-propositions is based on the identity of observable consequences, but it is not a rejection of the unobservable by means of an 'Occam's razor'. The existing opinion that there are no straight lines in an Einsteinian Universe but only curved ones is not so exact in essence. It really speaks of the indiscernibility of 'straight' and 'curved' from a certain fixed point of view. We can recall here the well known considerations of Albert Einstein about an elevator and the principle of equivalence in general.

Perhaps the most interesting examples of such 'in principle non-observable' assertions are those which have the same theoretical (and empirical) sense but are different in sense from the 'usual' view point, from the point of view of common sense. They allow us to understand how we restrict in a theory our usual intuitive notions, even notions concerning what is identical and what is different from a theoretical point of view. That is, the theory describes the true scales of identity and differences. From the empirical viewpoint we see only long and short tracks of particles, and common sense says that the particles lived for different lengths of time. However, only the theory tells us that *really* the two particles lived for the same time but this time was *slower* in the case of one of them. It means that only a theory proved by facts tells us something about external reality, and our task, more and more complex, consists in the correct understanding of the sense of this theory.

Academy of Sciences of the Ukrainian SSR, Kiev

RYSZARD WÓJCICKI

TOWARDS A GENERAL SEMANTICS OF EMPIRICAL THEORIES

0. The printed version of this paper differs only slightly from the original one. In particular I decided to keep unchanged the rather loose and informal language of the version prepared for the purpose of oral presentation. This less rigorous way of dealing with the issues touched upon here seems to possess a valuable advantage: it allows one to expose the main ideas in a more straightforward and easy way. Though some parts of the paper do not match the standard of full preciseness, still I hope that the more important points are discussed accurately enough for the reader familiar with basic logical techniques to understand them correctly and, eventually, couch them in precise terms.

1. All the issues I am going to deal with are motivated by the following more general question. Is it possible to examine the relationships holding between an empirical theory θ and the phenomena to which θ is supposed to be applicable by means of substantially the same tools which are used within logical semantics, and how eventually can this be done? The problem is far from trivial. Empirical phenomena do not resemble the set-theoretic structures which serve as possible referents of theories subjected to metamathematical inquiry. What we can do, at best, is to represent empirical phenomena by structures of an appropriate sort. It is obvious, however, that any attempt to represent something which belongs to the real world by an abstract mathematical construction involves inevitably some idealizations which, besides being more or less far going, may also go in different directions. In applying abstract concepts in order to grasp properties of real things, we have to be aware that we can usually do this in many different ways, thus obtaining more or less satisfactory solutions to the problems we are interested in.

Butts and Hintikka (eds.), Basic Problems in Methodology and Linguistics, 31–39.
Copyright © 1977 by D. Reidel Publishing Company, Dordrecht-Holland. All Rights Reserved.

Let (P, u) be a pair consisting of a phenomenon P and a set u of some empirical methods by means of which one may gather empirical data on P relevant to θ. As a matter of fact, it is this pair (P, u) which can be represented by a set of set-theoretic structures rather than P alone. Call any such pair an *application* of θ. Knowing P we know where θ is applied; knowing u we know how it is applied in this particular domain. Clearly in order for (P, u) to be an *intended application*, P should be a phenomenon which is to be described by θ, and u should consist of methods which experts in the particular field of science to which θ belongs are ready to accept as legitimate. Neither of these requirements can be stated in the form of a strict, formal assumption, however. As a very weak counterpart of the requirement of legitimacy of the methods in u, we shall postulate their *coherence*: given any intended application (P, u) we shall assume that the methods in u are such that they never provide us with a logically inconsistent set of data. We shall assume even more. Let (P_t, u_t), $t \in T$ be all intended applications of θ, and let E_t be the set of all those hypotheses on P_t whose truth can be established by means of methods in u_t, call them u_t-*true* hypotheses, for short. The sets E_t need not be logically independent of each other, they may even simply overlap. Roughly speaking, we shall require the set-theoretic union $\bigcup E_t$ of all E_t to be logically inconsistent, which obviously implies the coherence of all u_t. Clearly, the converse need not be true: $\bigcup E_t$ may be inconsistent even if all u_t are coherent. Observe that the requirement of the logical inconsistency of $\bigcup E_t$ binds different applications of θ and thus plays the same role as some requirements stated in terms of constraints play in [7].

In general, the language in which the empirical data available in a particular application (P, u) of θ can be stated may differ essentially from that of θ. I shall, however, confine myself to the much more simple situation, in the case of which the empirical data available in (P, u) can be couched in the language of θ augmented, eventually, by a number of individual symbols of the same semantic category as some of the variables appearing in this language. Such an augmentation may be necessary in order to have at our disposal names of the objects to which the data established by means of u refer, for example the names of objects on which some quantities are measured.

It should be emphasized that, in general, in order to define the set u it is not enough to define simply an appropriate set of measurement techniques. Often, we have to define, in addition, the meaning which some of the symbols (terms) involved in θ acquire under the application (P, u). For example, if θ deals with positions of mass-points but in the application (P, u) the notion of a position refers to physical bodies then, apart from selecting a length unit and a suitable frame of reference, we have to make clear what exactly is meant by the position of a body. We may identify it, for example, with the position of the mass centre of that body or, perhaps with the position of a point on its surface or select still other stipulations. Any such decision should be motivated by some theoretical considerations; incidentally, this calls our attention to the fact that empirical methods are theory dependent.

2. Let me discuss now, in possibly concise manner, what set-theoretic structures can be assigned to applications of an empirical theory. To begin with, assume that (P, u) is an application such that all hypotheses about P decidable by means of methods in u, call them u-*decidable*, are of the form:

(1) $$F_i(t, a_1, ..., a_k) = x \pm \varepsilon,$$

where $i = 1, ..., n$, x, and ε are real numbers and $\varepsilon > 0$. The symbol t represents a time instance, $a_1, ..., a_k$ stand for the objects on which the quantity F_i is measured. Applying a more pedantic notation we should write $k(i)$ rather than simply k. Assuming that for each formula of the form (1) there is a least ε for which this formula is still u-decidable, and taking into account that u is assumed to be coherent, we may define a function F_i^u corresponding to F_i by postulating that the identity:

(2) $$F_i^u(t, a_1, ..., a_k) = [x - \varepsilon, x + \varepsilon]$$

holds true whenever the corresponding formula (1) is u-true. The function F_i^u will be said to be an *operational counterpart* of F_i. Observe that the values of F_i^u are real number intervals.

The sequence:

(3) $(F_i^u, ..., F_n^u)$

is exactly what we have been looking for: a set-theoretic representation of (P, u). A sequence of the form (3) will be said to be an *operational structure* defined by (P, u).

Let us define a notion which will be useful later. Given two applications (P, u), (P, v) which differ only with respect to the sets of empirical methods they involve, we shall say that (P, u) is *more refined* than (P, v) if and only if for each sequence $t, a_1, ..., a_k$ of an appropriate sort, and for each $i = 1, ..., n$, the interval which F_i^u assigns to that sequence of arguments is a subinterval of the interval assigned to it by F_i^v.

3. Let me split the further discussion between two specific cases. First, assume that apart from $F_1, ..., F_n$ there are no symbols in the language of θ which have to be provided with an empirical interpretation. Thus all the remaining symbols involved in θ are either logical or mathematical. Under this assumption I shall say that θ is *confirmed* by the application (P, u) or, equivalently, that it is *approximately true* in the operational structure defined by (P, u) if and only if among all idealizations of the operational structure (3) there is at least one which determines a model (in the logical sense of the word) of θ. By an *idealization* of the sequence (3) I mean any sequence

(4) $(G_1, ..., G_n)$

of real valued functions, each G_i being defined for exactly the same arguments as the corresponding F_i^u and moreover taking values such that for all suitable arguments $t, a_1, ..., a_k$ the following holds:

(5) $G_i(t, a_1, ..., a_k) \in F_i^u(t, a_1, ..., a_k)$.

For further discussion of the notion of (approximate) truth defined by means of the notion of an operational structure cf. [1], [8], [9].

Applying the method of expanding a model defined for the empirical part of a language onto the theoretical part of that language (cf. [5], [6], [7]) we may handle the case when the symbols $F_1, ..., F_n$ involved in the application (P, u) do not exhaust all non-logical and non-mathematical terms of θ. In this case, we shall say that (P, u) *confirms* θ or, equivalently, that θ is *approximately true* in the operational structure defined by (P, u) if and only if among all idealizations of this structure there is one which can be expanded to a model of θ.

4. Let me discuss now a specific and more sophisticated situation which we face in some theories, especially in quantum mechanics. The u-decidability of each of the sentences $h_1, ..., h_m$ need not guarantee that they are *u-codecidable:* applying the methods u, say to h_1 we may lose the possibility of applying them to h_2 and vice-versa. Assuming that for each finite set $h_1, ..., h_m$ of hypotheses on P it is known whether this set is u-codecidable or not (which in fact amounts to assuming that the methods in u are well defined) we shall define an arbitrary set X of u-decidable sentences to be *u-codecidable* if and only if each finite subset of X is u-codecidable. Applying Zorn's lemma we may prove, by an easy argument, that among all u-codecidable sets of u-true sentences there are maximal ones. To each such set we may assign again an operational structure in substantially the same manner as that which we applied in assigning an operational structure to the whole set of u-true sentences. There are some points which call for clarification here, but I shall deliberately ignore them. The operational structures corresponding to maximal u-codecidable sets will be said to be the *complementary* structures defined by (P, u). Extending the notion of confirmability and thus also the notion of approximate truth to the case being dealt with now, we shall say that θ is *confirmed* by (P, u) (θ is *approximately true* in the set of complementary structures defined by (P, u)) if and only if θ is approximately true in each structure of that set.

To complete the discussion we have carried out, one case more should be taken into account. Call a method in u *dispersive* (cf. [2]) if applying it to a hypothesis h, we cannot establish whether h or not h, but only what is the probability (objective or perhaps subjective) that h. The prevailing majority of the empirical methods applied in science fall under this category. In spite of its obvious importance, I am afraid, however, that I would not be able to discuss the dispersive case. The problems it raises are far too involved to be dealt with in as short and as loose a way as those I have touched upon so far.

5. In evaluating theories, besides being interested in whether they are confirmed by their intended applications, we want also to know how strongly they are confirmed. A purely tautological theory, if anyone had proposed such, would be confirmed by any application whatsoever, having at the same time no practical value.

The notion of confirmation has been defined here in a way deviating from the familiar ones. The point is that in the definitions proposed in Sections 3 and 4 this notion is related to an application (P, u) as a whole rather than to a particular test carried out within the application. This last notion can be defined as follows. A pair (E, e) composed of a set of hypotheses E and a single hypothesis e will be said to be a *test* in (P, u) if and only if the union $E \cup \{e\}$ is u-codecidable and each element of this union is a u-true hypothesis about P.

Given any such test (E, e) we shall say that:

(i) (E, e) *confirms* θ if and only if the following three conditions are satisfied:

(6) $\theta \cup E \vDash e,$

(7) $E \nvDash e,$

(8) $\theta \nvDash e,$

where $\vDash$ stands for semantic entailment, and $\nvDash$ indicates that the entailment does not take place.

(ii) (E, e) *disconfirms* θ if and only if $\theta \cup E \vDash \text{non-}e.$

(iii) (E , e) is *irrelevant* to θ if and only if it neither confirms nor disconfirms θ.

By an obvious argument one can prove that θ is confirmed by an application (P, u) if and only if no test in (P, u) disconfirms θ. This assertion may give rise to some criticism. One may feel that if a theory is confirmed by (P, u) then at least one test in (P, u) should confirm the theory; to put it in other words: one may object to calling the theory confirmed by (P, u) if all tests in (P, u) are irrelevant to it. I would grant this criticism, but at the same time the issue seems to me to be purely verbal. To discern between 'void' confirmation and 'real' confirmation we may, for instance, define a theory to be *positively confirmed* (not merely confirmed) by an application (P, u) if and only if at least some tests in (P, u) confirm θ and none disconfirm it. A purely tautological theory may serve as an example of a theory which being confirmed by all its "applications" is not positively confirmed by any one of them.

Given two theories θ and θ', which have at least some intended applications in common, we may say that θ is *at least as strongly confirmed* by an application (P, u) (common to the two theories) as θ' is if and only if both θ and θ' are confirmed by (P, u) and each test in (P, u) which confirms θ' confirms also θ. The stronger θ is confirmed by an application (P, u) the stronger the connections it establishes between empirical data concerning P available by u. Thus, if in all common applications θ is at least as strongly confirmed as θ' we might say that θ is *empirically stronger* than θ' or (in view of an obvious similarity between (6) and Hempel–Oppenheim's schema of the deductive-nomological explanation, cf. e.g. [3]) that the *explanatory power* of θ is not weaker than that of θ'. By producing a suitable example one may prove that two theories can be of the same empirical strengths though one of them logically stronger than other. The problems discussed now are closely related to the problem of factual content of empirical theories (cf. [10]).

6. Forming an empirical theory θ we always try to design it in such a way that it meets (among others) the following three requirements:

(i) The *range of applications* (i.e. the set of all phenomena to which θ can be applied) should be as wide as possible;

(ii) θ should be resistant to attempts to disconfirm it by empirical tests carried out within applications which are as refined as possible;

(iii) θ should be as empirically strong as possible.

These three requirements may be in conflict with each other. In particular quite often in order to strengthen a theory we have to narrow its range of applications. Usually we do this by dividing the range into certain subranges, and then developing, apart from the original general theory, some special theories related to the more specific sets of phenomena.

The possibilities óf improvement of a theory are always limited by its conceptual framework. As was convincingly argued by Kuhn [4], it often happens that after a period, sometimes a very long one, of successful development of the theory, we start realizing that to achieve further progress, and to solve the difficulties which the theory begins to encounter, we cannot help revising the whole conceptual apparatus of the theory in a radical and thorough way. Clearly by modifying the conceptual framework, we modify the theory as a whole: we have to reformulate the theorems (laws), at least so as to adjust them to the new concepts, we have to revise our conception of the range of the theory, and finally we have to rediscuss the assumptions on which the empirical procedures applied in the theory are based. One may get the impression that the new theory need not have too much in common with the old one. No matter however, how radical the changes are, the new theory should be confirmed by all applications (perhaps redefined in a suitable way) by which the old theory was confirmed. One may reasonably expect even more: the new theory should be confirmed by all empirical tests by which the old was confirmed. All these are only some preliminary and loose remarks however. I made them with the intention of showing how eventually the problem could be examined within the account of empirical theories I have tried to outline. Clearly, the decisive criterion of adequacy of this account consists in whether it can be successfully applied in dealing with the key problems of philosophy of science to which, no doubt, the problem of intertheory relations belongs.

BIBLIOGRAPHY

[1] Dalla Chiara Scabia, M. L. and Toraldo di Francia, F., 'A logical Analysis of Physical Theories', *Rivisita del Nuovo Cimento*, Series 2, **3** (1973), 1–20.
[2] Giles, R., 'A Non-classical Logic for Physics', *Studia Logica* **4** (1974), 397–415.
[3] Hempel, C. G., *Philosophy of Natural Science*, New York, 1963.
[4] Kuhn, T. S., *The Structure of Scientific Revolutions*, Chicago, 1970.
[5] Przełęcki, M., *The logic of Empirical Theories*, London, 1969.

[6] Przełęcki, M. and Wójcicki, R., 'The Problem of Analyticity', *Synthese* **19** (1969), 374–399.
[7] Sneed, J. D., *The Logical Structure of Mathematical Physics*, Dordrecht, 1971.
[8] Wójcicki, R., 'Basic Concepts of Formal Methodology of Empirical Sciences', *Ajatus Yearbook of the Philosophical Society of Finland* **35** (1973), 168–196.
[9] Wójcicki, R., 'Set Theoretic Representations of Empirical Phenomena', *Journal of Philosophical Logic* **3** (1974), 337–343.
[10] Wójcicki, R., 'The Factual Content of Empirical Theories', in *Rudolf Carnap, Logical Empiricist: Materials and Perspectives* (ed. by J. Hintikka), Dordrecht, 1975, pp. 95–122.

II

IDENTIFIABILITY PROBLEMS

HERBERT A. SIMON

IDENTIFIABILITY AND THE STATUS OF THEORETICAL TERMS

Much recent discussion about the structure of scientific theories has centered on the status of theoretical terms, that is, terms whose direct referents are not observables. It has been asked whether such terms are always *definable*, or if not definable, whether they are always *eliminable*. Another line of inquiry asks under what circumstances theoretical terms, or sentences containing them are *meaningful*; and still another asks under what circumstances they are *identifiable*, that is, can be estimated from observational data.

The present paper has two main objectives. The first is to clarify the relations among the four concepts mentioned above: definability, eliminability, meaningfulness, and identifiability. Since separate literatures have sprung up around each of them, it is important to observe that a single underlying theme can be found that threads through all of them. The second objective of the paper is to comment on why it is useful to employ theoretical terms in scientific theories, and how that usefulness relates to their identifiability.

Before we launch our discussion on either of these two topics, it is necessary to take care of some formal preliminaries, in particular, to say what is meant here by 'scientific theory'.

1. FORMAL PRELIMINARIES

The discussion in this paper will be conducted on a relatively informal level, since the fundamental issues that need to be raised and the basic distinctions that need to be made do not depend upon details of a formalism. When appropriate, reference will be made to the literature where some of these questions are treated more rigorously. Nor will the widest possible generality be sought, but rather, a sufficient generality to encompass a number of 'real life' examples of scientific practice.

Butts and Hintikka (eds.), *Basic Problems in Methodology and Linguistics*, 43–61.

As our formalism, we employ a first-order language, L, with equality, containing certain function symbols (possibly including constants). One or more of the function symbols, $O_1, ..., O_k$, is drawn from a set O, and one or more, $T_1, ..., T_l$, from a set T. There is an infinite set, X, of variable symbols, x_i. By a *theory*, F, in L, we will mean a set of formulas of L comprised of the logical axioms together with a formula, $F(O, T)$, which contains symbols from O and possibly T.

Next, we introduce a structure, $\langle D, O_M, T_M \rangle$, consisting of a set, D, of elements, and two sets of functions on D (call them O_M and T_M) corresponding to the two sets of function symbols, O and T, in L. The ranges of all functions will be either the real numbers, or the truth values, T and F. To each distinct element, d, of D is assigned a constant symbol to serve as its name.

The functions of O_M are to be interpreted as *observables* whose values can be determined in each instance by observing the appropriate set of objects at the appropriate times. The members of T_M, on the other hand, are *theoretical functions and relations*, whose values cannot be determined directly from observations. The structure, $\langle D, O_M \rangle$, obtained by removing the functions and predicates of T_M from $\langle D, O_M, T_M \rangle$ is called the *restriction* of the latter (Shoenfield, 1967).

DEFINITION. The *observable consequences*, $H(O)$, of a theory are the members of the class of all consequences of F that do not contain symbols from T.

Let $F_c(O, T)$ be a closed formula obtained from $F(O, T)$ by replacing each free variable in $F(O, T)$ by one of the constants that corresponds to an element of D. If all the $F_c(O, T)$ that can be constructed in this way are true of $\langle D, O_M, T_M \rangle$, then this structure is a model for the theory F.

In most of what follows, the elements of D will be 'observations', rather than 'objects', so that we will write, for example, $c(x)$ and $r(x)$ for the values of the current and resistance, respectively, of a particular electrical circuit, observed at some instant of time. For some purposes, we may specialize the set D to the product, $I \times J$, of sets, I, of observations ('instants') and, J, of objects. Then x_{ij}, for example, will represent object j observed at time i. If certain functions are assumed to be constant over observations, we may also refer to them by expressions like $m(x)$, for the mass of an object, x. When the contrary is not stated, however, the members of D should be interpreted as observations, so that extending the set, D, means taking additional observations.

DEFINITION.   For a given structure, $\langle D, O_M \rangle$, iff there exists a T_M on D, and a corresponding T in L, such that $F(O, T)$ is valid in $\langle D, O_M, T_M \rangle$ then $\langle D, O_M \rangle$ is *expandable to a model* for the theory F.

DEFINITION.   Consider a model, m_k, and a second model, m_{k+}, obtained by annexing (zero or more) elements to D so as to obtain D', $D \subseteq D'$, and extending the O_M's to D'. We will call the new model, m_{k+}, an *extension of* m_k, and will write $m_{k+} \geqslant m_k$, and if $D' \supset D$, $m_{k+} > m_k$.

Expanding a model is to be interpreted as introducing new theoretical terms into it, while extending a model is to be interpreted as taking additional observations.

Some conditions will now be introduced to limit the concept of 'theory' so that anything we call a well-formed theory will be operational and testable (Simon and Groen, 1973). Let M be a set of structures, $\langle D, O, T \rangle$. Then we will designate by $\mathcal{M}_*$ the restriction, $\langle D, O \rangle$, obtained from M by deleting the T predicates. Thus, to each member of $\mathcal{M}_*$ there will correspond a subset of M.

In M consider the set of all models for $F(O, T)$, and call this set M^T. Restrict each member of M^T, obtaining a set, $\mathcal{M}_*^T$, of models in $\mathcal{M}_*$. Clearly, all members of $\mathcal{M}_*^T$ are expandable to models for $F(O, T)$. Next, consider the set of all models in M for $H(O)$. Call this set M^O. Again, remove the T relations to obtain a set $\mathcal{M}_*^O$, of models in $\mathcal{M}_*$. Again, all members of $\mathcal{M}_*^O$ are expandable to models for $H(O)$. Moreover, by their construction, the members of $\mathcal{M}_*^O$ are all models for $H(O)$ in $\mathcal{M}_*$.

Testability of a theory would seem to require that two conditions be satisfied:

(1)   If the theory is false, there should exist one or more *finite* sets of possible observations that would disconfirm it.

(2)   If the theory is disconfirmed by some set of observations, m_k, it should not be possible to make additional observations of such a kind that the theory is now consistent with the extended set, $m_{k+} > m_k$. More formally:

DEFINITION.   A theory, $F(T, O)$, is *finitely testable* iff $\exists m(m \notin \mathcal{M}_*^T)$ and $\forall m[(m \notin \mathcal{M}_*^T) \rightarrow \exists m_k((m_k \leq m) \wedge (m_k \notin \mathcal{M}_*^T))]$, for finite k.

DEFINITION.   A theory, $F(T, O)$, is *irrevocably testable* iff $\forall m[\exists m_k((m_k \leq m) \wedge (m_k \notin \mathcal{M}_*^T)) \rightarrow (m \notin \mathcal{M}_*^T)1]$.

DEFINITION.   A theory, $F(T, O)$ is *FIT* iff it is both finitely and irrevocably testable.

Simon and Groen (1973, pp. 371–5) have set forth a number of reasons for insisting that all scientific theories be *FIT*. This view, however, is by no means universally accepted (e.g. Sneed, 1971, pp. 57–59; Tuomela, 1973, pp. 63–64). One important objection to it is that theories incorporating certain standard mathematical axioms may not be FIT. For example, the assertion that a certain set of magnitudes is Archimedean is not finitely and irrevocably testable (see Adams *et al.*, 1970; Adams, 1974, pp. 439–444), nor are axioms of continuity, for even if a model failed to satisfy them, no finite set of observations could reveal this failure.

Adams *et al.* (1970) introduce the important notion of *data equivalence*. Two theories, T and T', are data equivalent iff they are consistent with exactly the same finite sets of observations. An axiom is 'technical' if removing it from a theory produces a new theory that is data equivalent to the original one. 'Technical' axioms, then, contribute nothing to the empirical (finitely testable) content of the theory. Adams *et al.*, observe that the Archimedean axiom and continuity axioms are generally technical in this sense.

The question of whether theories that are not FIT should be admitted into consideration might therefore be reformulated as a question of what useful function is performed by 'technical' axioms. We will not pursue that question here, but will limit the following discussion to FIT theories.

2. Eliminability of Theoretical Terms

The topic of eliminability can be treated briefly by reference to the results reported by Simon and Groen (1973). The theoretical terms of a theory are *eliminable* (more precisely, *Ramsey-eliminable*) if, in a certain sense, all of the empirical content of the theory can be stated in terms of the observable relations, O. More formally:

DEFINITION.   If, for fixed D, $\mathcal{M}_*^T = \mathcal{M}_*^O$, that is, if all models in $\mathcal{M}_*$ for $H(O)$ are expandable to models for $F(O, T)$, then we will say that the theoretical relations, T, are *D-Ramsey-eliminable*. If the relations, T, are *D-Ramsey-eliminable* for all D, then we say, simply, that they are *Ramsey-eliminable*.

Simon and Groen proved (their Theorem 3, p. 375) that if a theory, $F(O, T)$, is *FIT*, i.e. finitely and irrevocably testable, then its theoretical terms, T, are Ramsey-eliminable.

This result tells us that the theoretical terms of a FIT theory are never indispensable, hence, if we wish to retain them it must be for reasons of convenience rather than necessity. To see what is involved in their elimination, it will be useful to look at a concrete example of a physical theory: an axiomatization of Ohm's Law, first proposed in Simon (1970). The theory is axiomatized by defining a set-theoretical predicate, *Ohmic circuit*, as follows:

T1. Γ is a system of *Ohmic observations* iff there exist D, r, c, such that:

    (1) $\Gamma = \langle D, r, c \rangle$;
    (2) D is a non-empty set;
    (3) r and c are functions from D into the real numbers;
    (4) for all $x \in D$, $r(x) > 0$ and $c(x) > 0$.

Γ' is an *Ohmic circuit* iff there exist D, r, c, b, and v such that:

    (5) $\Gamma' = \langle D, r, c, b, v \rangle$;
    (6) $\Gamma = \langle D, r, c \rangle$ is a system of Ohmic observations;
    (7) v and b are real numbers;
    (8) for all $x \in D$,

    (a) $c(x) = v/(b + r(x))$

In this system, r and c are to be interpreted as observables; b and v as theoretical terms. This theory is finitely testable, since, in any structure that is not a model, there exist substructures, m_3, with values of $\{c_i(x), r_i(x)\}$ that do not satisfy (a). It is irrevocably testable, for a set $\{c_i(x), r_i(x)\}$ will not satisfy (a) if any of its subsets fail to do so. In fact, it is k-testable, with $k = 3$.

Suppose, now, that three pairs of independent observations, c_1, r_1, c_2, r_2, c_3, r_3, are made of c and r. We can use the first two pairs of observations to derive from (a) a pair of simultaneous equations that we can solve for v and b.

$$b = (c_2 r_2 - c_1 r_1)/(c_1 - c_2)$$
$$v = c_1 c_2 (r_2 - r_1)/(c_1 - c_2)$$

If we now substitute these values for v and b in (a), and substitute also c_3 and r_3 for $c(x)$ and $r(x)$, respectively, we obtain a relation among three pairs of observations. But since these could be any three pairs of observations, we may write the relation, generally, thus:

(b) $c_i c_j (r_i - r_j) + c_i c_k (r_k - r_i) + c_j c_k (r_j - r_k) = 0.$

Now the members of $H(O)$, the observable consequences of $F(T, O)$, are all derivable from (b), which contains no members of T. In fact, the theory with the theoretical terms eliminated could be axiomatized by (1)–(4) together with (b).

What is the difference between the two axiomatizations (other than the absence of members of T from the second)? In some intuitive sense, (a) is a more 'compact' and parsimonious expression than (b). In particular, (a) makes reference to only a single observation, $\{c(x), r(x)\}$, while (b) states a relation among a triplet of observations $\{c_i, r_i, c_j, r_j, c_k, r_k\}$. Finally, a point that will be taken up again later, (a) 'reifies' the voltage and internal resistance of the battery, while the battery and its properties are not mentioned at all in (b).

Hintikka (1965, 1970) and Hintikka and Tuomela (1970) have proposed a distinction between 'depth' information and 'surface' information that is illustrated by the two axiomatizations of Ohm's Law. Specifically, Hintikka (1965) has suggested 'that the intuitive meaning of the concept of depth of a sentence, s, is the number of individuals considered together (in their relation to each other) in the deepest part of s'. Translating 'individuals' as 'observations', sentence (b), without theoretical terms, is of depth 3, while sentence (a), containing such terms, is of depth 1. The depth of a theory may also be measured by the largest number of layers of nested quantifiers that appears in the sentence formalizing the theory (Tuomela, 1973, p. 46). Applying this measure to a first-order formalization of the Ohm's law axiomatization, Tuomela (1973, pp. 161–162) finds the depth of (b) to be 15 and the depth of (a) to be 5, again showing a depth reduction by a factor of three in (a) as compared with (b). Hintikka and Tuomela (1970, p. 310) regard the reduction in depth as a main advantage derivable from introducing theoretical terms into a theory.

Of course this relation between depth and the presence or absence of theoretical terms is not peculiar to the example of Ohm's Law. In general, if we have a sentence containing such terms, then, given a sufficient number of observations, we may be able to solve for the values of the theoretical terms in terms of the values of observables. In this way we replace the sentence by a new (and deeper) one from which the theoretical terms have been eliminated. Roughly, we would expect the sentence to increase by one in depth (according to the first measure mentioned above) for each theoretical term eliminated. This elimination process will be discussed more fully later under the topic of identifiability.

3. THE MEANINGFULNESS OF THEORETICAL TERMS

The concept of meaningfulness of theoretical terms to be proposed here is closely related to some ideas of Przełęcki (1974, p. 347). The basic notion is very simple: for a term to be meaningful, some sentences containing it (and containing it 'essentially') must make empirical statements about the world. Przełęcki limits her discussion to situations where the theoretical terms, T, are introduced by meaning postulates that are non-creative relative to the postulates that contain only O-terms. In the Ohm's Law example of the previous section, this condition is not met, for the T-terms are introduced there by (a), which is creative relative to the other postulates. In the present discussion, it is not assumed that the postulates introducing the T-terms are non-creative. There seems to be no real basis for the traditional insistence that definitions be separated sharply from empirical laws (Simon, 1970), and such a separation appears to be the exception rather than the rule in usual formulations of scientific laws.

Let us assume that in a theory, $F(O, T)$, all of the T-terms are constant functions, as they are in the axiomatization of Ohm's Law that we have been using as an example. For fixed values of the T-terms, there are, as before, models, M^T, for the theory, as well as a set of models, $\mathcal{M}_*^T$, obtained by restricting each member of M^T by removing the T-terms.

Now consider a whole family of theories, obtained by taking different values for the T-terms of F. Associated with each theory of this family will be a set of models and a set of (possibly overlapping) restricted models. Consider the set union of the sets of restricted models, $\mathcal{M}_*^T$, calling it $\mathcal{M}_*$. Let $R(T)$ be the function whose domain consists of the admissible sets of values of the T-terms, and whose range consists of the corresponding subsets of $\mathcal{M}_*$. If there exists some member of the domain of R whose value is a proper subset of $\mathcal{M}_*$, then the T-terms are *meaningful*. Let S be a sentence that asserts that the T-terms lie in such a region. Then knowledge that S is true provides information about the possible values of the O-terms. Hence the sentence, S, may be termed meaningful.

These definitions attribute meaningfulness to the set of T-terms as a whole. For some purposes, we may wish to attribute meaningfulness or meaninglessness to particular members of that set. In the Ohm's Law example, we might want to ask separately whether v is meaningful and whether b is meaningful, rather than simply whether the pair, $\{v, b\}$ is meaningful. The following definition meets this need:

Let $F(O, T)$ be a family of theories, as above. Construct a subfamily of F by assigning particular values to all of the T-terms except t_k, say. Then, relative to this assignment, we can define the meaningfulness of t_k exactly as before. Moreover, we can introduce a weaker concept of meaningfulness: the term, t_k, is meaningful if it is meaningful for some assignment of the remaining T-terms.

The central idea of all of these definitions is that, for a theoretical term to be meaningful, knowledge of the value of that term must imply some limitation on the range of the O-terms. All sorts of weaker or stronger concepts of meaningfulness may be introduced that share this general underlying idea.

The concept of meaningfulness induces a reciprocal relation between the T-terms, on the one hand, and the O-terms on the other. For consider the space, $O \times T$, which is the product space of the ranges of the O-terms and T-terms, respectively. Suppose that, in a given theory, it is true that $t \in T'$ implies $o \in O'$, where T' and O' are proper subspaces of T and O, respectively. Then it follows that $o \in C(O')$ implies $t \in C(T')$, where $C(x)$ is the set complement of x. Hence, information about the O-terms implies a restriction on the possible values of the T-terms.

A definition of meaningfulness could be based upon this implication instead of the inverse implication that has been used here. In fact, the former is the path followed by Przełęcki (1974), and earlier by Suppes (1959). In particular, Przełęcki's criterion of weak meaningfulness is roughly equivalent to the requirement that $o \in O'$ implies $t \in T'$, where O' and T' are proper subspaces of O and T, respectively.

The procedure followed in the present paper seems, however, to correspond more closely to the everyday intuitive meaning of the phrase 'empirically meaningful' than does the proposal mentioned in the last paragraph. It appears to be more natural to say that a term is meaningful if it conveys information about the observable world than to say that it is meaningful if something can be inferred about its value from information about the real world. In fact, as we shall see in the next section, the latter idea would seem closer to the concept of 'identifiability' than to the concept of 'meaningfulness'.

4. THE IDENTIFIABILITY OF THEORETICAL TERMS

From the discussion of the previous section it is apparent that identifiability is, in a certain sense, a converse of meaningfulness. Theoretical terms are meaningful to the extent that their values restrict the values of the observables; they are identifiable to the extent that their values are restricted by specification of the values of the observables.

Moreover, by the law of contraposition, the relation between the two terms is always reciprocal, as we saw above. For if restriction of o to some subspace of O restricts t to some proper subspace of T, then restriction of t to the complementary subspace of T restricts o to the complementary subspace of O.

Identifiability is a matter of degree, and a number of weaker and stronger forms of identifiability have been defined (Simon, 1966, 1970; Tuomela, 1968). By 'strength' is meant *how much* restriction is placed upon the possible values of T by a given amount of information about O. Since the topic has been treated elsewhere (see Tuomela, 1968, and the paper by Sadovsky and Smirnov in this symposium), only a few comments will be made here about weak and strong identifiability.

Perhaps the most important consideration in selecting a definition of identifiability is that axiomatizations of theories may sometimes be simplified if complete identifiability of the theoretical terms is not insisted upon. For example, the terms, v and b, in the axiomatization of Ohm's Law cannot be estimated from the values of observables unless there are at least two distinct observations. An additional axiom can be added that imposes this requirement; but alternatively, it may be satisfactory to define identifiability to mean 'unique estimatability provided there are a sufficient number of independent observations'. 'Sufficient number' here will generally mean one less than the number that measures the depth of the sentence obtained by eliminating the theoretical terms. (This number is closely related to the usual notion of *degrees of freedom.*)

A closely related issue arises when the theoretical terms of a system are estimatable from the observables except for certain special cases. For example, in a system of Newtonian particle mechanics, the mass ratios of two particles comprising an isolated system can be estimated from observations of positions and velocities at a sufficient number of distinct points in time, unless each of the particles is itself an isolated system. Concepts of *almost everywhere definability* have been proposed in order to rule out unidentifiability arising from such special circumstances (Simon, 1966, 1970).

Finally, *definability*, in the sense of Tarski, is simply the strongest form of identifiability, which admits none of the qualifications discussed in the two previous paragraphs.

While identifiability and definability are thus closely related concepts, neither has anything to do with eliminability, for the theoretical terms of a FIT theory, although always eliminable, need be neither definable nor identifiable. A simple and frequently cited example will be instructive. Consider a market for a single commodity, such that each observation consists of a quantity, $q(x)$, and a price, $p(x)$, at which the quantity is sold and bought. A theory of this market might consist of the two equations:

(c) $q(x) = a_1 p(x) + b_1$

(d) $q(x) = a_2 p(x) + b_2$

Equation (c) is the supply relation, which expresses the quantity offered in the market as a function of the price; while (d) is the demand relation, which expresses the quantity that will be purchased by buyers as a function of the price. The a's and b's are theoretical terms that define the hypothetical supply and demand curves. These two equations, taken together, are equivalent to the assertion that $p(x) - p_*$ and $q(x) = q_*$ are constants, independent of x. Since there are only two equations, but four theoretical terms, it is obvious that the latter are not identifiable.

This example also throws further light on the reciprocal relation between identifiability and meaningfulness – shows, in fact, that this relation need not be symmetric. As a set, the theoretical terms are meaningful, for assigning values to them determines the values of p_* and q_*. If we hold any three of the theoretical terms constant, then, by our earlier definition, the fourth is meaningful, because the admissible values of p_* and q_* then depend on its value. In the opposite direction, however, knowledge of the values of p_* and q_* does not identify any of the theoretical terms, but only two functions of them (obtained by solving (c) and (d) for the observables). Some general results of this kind are derived formally by Rantala in his paper for this symposium.

5. IDENTIFIABILITY AND CHANGE OF STRUCTURE

It is well known, and often pointed out in the econometric literature (Koopmans, 1953), that if a theory is going to be used only to predict values of the observables, then it is of no significance whether or not the theoretical terms are identifiable, provided that they remain unchanged. Thus, in the simple market example of the last section, a single observation determines the price and quantity of the good that will be exchanged, and if the theory is correct, the same price and quantity will hold for all future observations. Unidentifiability of the T-terms does not interfere with the prediction.

If some change takes place in the world, however, altering the unobservable values of one or more of the theoretical terms, then the new equilibrium values of p and q will not be predictable until they are observed. Hence the possibility of prediction under such a change (called a change in *structure*) hinges on having some additional source of information about the values of the theoretical terms.

Suppose we know from some independent source that changes are taking place in the coefficients of the first (supply) equation, but not in the second (demand) equation. We might know, for example, that changes in weather cause large changes in production that are not predictable by the producers, hence cause large random shifts in supply. Then the second equation would continue to hold, but not the first, and we would observe values of p and q from which we could estimate a_2 and b_2. Similarly, if we knew that changes were taking place in the coefficients of the second equation but not the first, we could take observations to estimate the coefficients of the supply equation. The introduction of random shocks into one of the two mechanisms brings about identifiability of the coefficients of the other. If, however, random shocks are imposed on *both* mechanisms, then in order to achieve identifiability we must have information about the relations of the two sets of shocks, e.g. that they are uncorrelated.

This example shows that the individual equations that go to make up a theory may be interpreted as representing separate particular components of the system that the theory describes. We may have information about the system, not incorporated in the values of the observable functions, that enables us to identify theoretical terms that are otherwise unidentifiable.

More generally, a complex system having only a small number of observables but a large number of theoretical terms will leave most or all of the latter unidentifiable, unless the theoretical terms can be partitioned among components of the total system, and the behavior of each component observed independently of its relations with the rest of the system (i.e. independently of the remaining system equations). Partitioning of this kind is, of course, one of the basic ideas underlying the experimental method. Notice that the applicability of the method hinges on localizing particular theoretical terms to particular system components.

As a further illustration of these ideas, consider a system consisting of two Ohmic circuits, A and B. Suppose that, by making observations on them, we estimate v_A, b_A, v_B, and b_B. If, now, we interchange the batteries of the two circuits, inserting the battery of A in circuit B, and the battery of B in circuit A, we will be tempted to predict (and probably correctly) that the values of the v's and b's that we will now estimate by new observations will be similarly interchanged. To make this prediction, we are obviously using some additional knowledge that associates particular theoretical terms with particular components (in this case, the batteries) of the systems. What is the source of this knowledge?

The only observations of which the formal axiomatization speaks are the observations of the current and the resistance. But in point of fact, the experimenter can also observe certain of his own interventions in the system: his replacement of the resistance wire with a longer or shorter one, or his interchanging of the batteries between the two circuits. The formal system could be expanded to assert that the values of certain theoretical *or* observational terms will remain constant only if certain components of the system remain undisturbed. Even more powerful would be assertions that particular theoretical terms are associated with particular components of the system, and that this association remains unchanged when the components are reassembled in a new way.

Postulates of this kind are implicit whenever, for example, the weight of an object is assigned by placing it on a balance with objects of known weight, and then the weight so determined is used to make calculations in new weighings. In this case again, derived quantities are associated with particular parts of the system, and are assumed to travel with these parts. In fact, this assumption that weights remain invariant from one weighing to another is essential to make the ratios of the weights of a set of objects identifiable. The assumption is, in fact, implicit in the usual axiomatizations of fundamental measurement (for a simple example, see Sneed, 1971, pp. 18–19), where the members of the basic sets in terms of which the systems are described are objects rather than observations.

It was suggested earlier that it is sometimes convenient to factor the space of observations into a product space of objects and observations, $I \times J$. If the identity of particular objects can be maintained from one observation to the next, then theoretical terms that denote constant properties of those objects become functions only of the object and not of the observation, thus reducing greatly the number of independent theoretical terms, and enhancing correspondingly the chances of achieving identifiability. In classical particle mechanics, for example, inertial mass is just such a theoretical term.

It is easy to see how we can apply these ideas formally to the axiomatization of Ohm's Law. We introduce sets, B, W, and A, whose members correspond to particular batteries, wires and ammeters, respectively. Then r, $c(x)$, v and b become functions from W, $A \times D$, B and B, respectively, to the real numbers. This means that these quantities now depend upon the objects with which they are associated. (Note that the value of c depends upon the observation, x, as well as the ammeter, A, with which it is associated.) Now, associated with any given observation, x, is a member of A, $A(x)$, a member of W, $W(x)$, and a member of B, $B(x)$. Corresponding to (8), we now will have:

(8′)   For all $x \in D$,

(a′) $c(A(x), x) = v(B(x))/(b(B(x)) + r(W(x)))$

As before, c and r are to be regarded as observables, v and b as theoretical terms. However, we shall later have occasion to reconsider that classification.

The achievement of identifiability is facilitated also by knowledge that certain theoretical terms are identically equal to zero. If a theory consists of a set of n linear algebraic equations among n variables, then, in general, the coefficients of those equations will be unidentifiable, for there will be $n^2 + n$ such coefficients, and only n independent equations from which to estimate them, no matter how many observations are taken. Suppose it is known, however, that all but n of the coefficients are zero. Then the non-zero coefficients are, in general, identifiable. Whence might the information that made the identification possible be derived? If it were known that certain of the variables could not act directly upon each other, then the coefficients describing those particular interactions could be set equal to zero. For example, if the system of equations described a bridge truss, and certain nodes of the truss were not connected directly with certain other nodes by physical members, then the coefficients describing the forces with which one of those nodes acted upon the other would be zero. If the number of directly connected nodes were small enough, the stresses in the truss would be identifiable – it would form a so-called *statically determinate* structure.

The idea that runs through all of these examples is that the identifiability of theoretical terms of a system can be increased if the theory incorporates axioms (1) localizing particular theoretical terms in particular components of the system, and (2) associating particular experimental operations with corresponding observable alterations of specific components. These alterations are represented by associating variables, both observables and theoretical terms, with system components.

Let us consider an experimental program for verifying and extending Ohm's Law. In our axiomatization of that law, we treated resistance and current as observables. Of course they are only relatively so, for they are defined, in turn, in terms of more directly observable quantities. In the present discussion, we wish to take that indirectness into account. We begin, as Ohm did, with a simple circuit and two observables: the readings on the angular displacement of a magnet, and the length of a wire of specified composition and diameter inserted in the circuit. The first observable is used to define the theoretical term 'current', the second to define the theoretical term 'resistance'. Now, by inserting resistances of different lengths in the circuit, we obtain a number of independent observations of the current, and can estimate the voltage of the battery. Having done so, we can next experiment with wires of constant length but different cross-section. The term 'resistance' was originally defined only for wires of a specific cross-section, but now that the voltage and internal resistance have been estimated, we can use their values, together with the observed values of the current, to calculate the values of the new resistances. In this way we can extend the theoretical term, 'resistance', to apply to conductors of any shape or dimension or material that we can insert in the circuit. We can employ the same general method, now using known voltages and resistances, to calibrate different current-measuring devices, and in this way to extend the definition of the theoretical term 'current'.

In terms of our expanded formalism, using the sets W, A, and B, this experimental program can be interpreted as starting with r and c for subsets of W and A, respectively, as observables, and the values of r and c for the remaining members of W and A, as well as the values of v and b for all members of B as theoretical terms. Knowledge of the values of the initial set of observables is sufficient to enable us to 'bootstrap' estimates of all of the theoretical terms by performing observations on appropriate circuits.

Once we are able to assign definite voltages and resistances to physical components that can be used to construct current-bearing systems, we can now also begin to determine the laws of composition of such components when they are organized in more complex arrangements. We can add axioms to the system, for example, for the behavior of two resistances in series or parallel, or of two batteries in series or parallel. Thus, by a procedure rather analogous to analytic continuation in the theory of complex variables, we are able to extend the domain of identifiability of theoretical terms from a small initial region to a wider and wider range of increasingly complex situations.

The comment was made in an earlier section of this paper that to identify a theoretical term is to 'reify' it. The meaning of that comment should now be clear. Even though the theoretical terms are not directly observable, if they can be associated, as invariant properties, with particular physical objects, and if these objects can be made to interact in various configurations with each other, then a high degree of identifiability may be attained even in complex systems. There is no magic in this process. The assumption that a theoretical term is an invariant property holds down the number of such terms whose values have to be estimated, while the ability to recombine the system components in different configurations increases the number of equations that they must, simultaneously, satisfy. Of course the axiomatization of a system that permits such 'analytic continuation' as a means for identifying theoretical terms must include appropriate composition laws for the configurations of components that are to be examined.

The developments proposed in this section can be interpreted as a possible answer to Sneed's (1971, p. 65) call for 'a rationale for the practice of exploiting values of theoretical functions obtained in one application of [a] theory to draw conclusions about other applications of the theory'. The requirement that particular theoretical terms be associated with specific objects, and that they retain their values when reassembled in different subsystems, would seem to correspond closely with Sneed's concept of a 'constraint' to be applied to a theoretical term when it is moved from one application to another.

6. Conclusion

This paper has undertaken to explore the relations among the concepts of 'eliminability', 'meaningfulness', 'identifiability', and 'definability' as applied to theoretical terms in axiomatized theories. Theoretical terms are always eliminable from finite and irrevocably testable theories. Hence, they are retained for reasons of convenience, and not because they are indispensable. One such reason is that the statement of the theory requires 'deeper' sentences, in the sense of Hintikka, if the theoretical terms are eliminated than if they are retained. Another reason for retaining theoretical terms is that, by attributing them to particular physical components of systems, we can conveniently bring a whole range of partial theories within a single general formulation, gaining identifiability in the process.

Meaningfulness and identifiability have been shown to be opposite sides of the same coin, though a coin that is not necessarily symmetrical. Meaningfulness involves predicting from known values of theoretical terms to values of observables; identifiability involves making estimates from values of observables to the values of theoretical terms. Both meaningfulness and identifiability are matters of degree – stronger and weaker notions of either can be introduced, and definability, in the sense of Tarski, can simply be regarded as the strongest form of identifiability. Meaningfulness of a set of theoretical terms always implies some degree of identifiability, in at least a weak sense, and vice versa.

In the final section of the paper it was argued that our knowledge about a system is usually not restricted to knowledge of the values of observables. In addition, we can label particular system components, and trace their continuity through operations that change the system configuration. Incorporating in the axiomatization postulates that associate specific theoretical terms with specific components, and that assert the invariance of the values of those terms is a powerful means for enhancing the identifiability of theoretical terms.

Carnegie-Mellon University

BIBLIOGRAPHY

Adams, E. W.: 1974, 'Model-Theoretic Aspects of Fundamental Measurement Theory', in L. Henkin (ed.), *Proceedings of the Tarski Symposium* (*Proceedings of Symposia in Pure Mathematics*, Vol. 25). American Mathematical Society, Providence, R.I., pp. 437–446.

Adams, E. W., Fagot, R. F., and Robinson, R. E.: 1970, 'On the Empirical Status of Axioms in Theories of Fundamental Measurement', *Journal of Mathematical Psychology* 7, 379–409.

Hintikka, K. J. J.: 1965, 'Are Logical Truths Analytic?', *Philosophical Review* 74, 178–203.

Hintikka, K. J. J.: 1970, 'Surface Information and Depth Information', in K. J. J. Hintikka and P. Suppes (eds.), *Information and Inference*, D. Reidel Publishing Company, Dordrecht, Holland, pp. 263–297.

Hintikka, K. J. J., and Tuomela, R.: 1970, 'Towards a General Theory of Auxiliary Concepts and Definability in First-Order Theories', in K. J. J. Hintikka and P. Suppes (eds.), *Information and Inference*, D. Reidel Publishing Company, Dordrecht, Holland, pp. 298–330.

Koopmans, T.: 1953, 'Identification Problems in Economic Model Construction', in W. Hood and T. Koopmans (eds.), *Studies in Econometric Method*, John Wiley & Sons, New York, pp. 27–48.

Przełęcki, M.: 1974, 'Empirical Meaningfulness of Quantitative Statements', *Synthese* 26, 344–355.

Shoenfield, J. R.: 1967, *Mathematical Logic*, Addison-Wesley Publishing Company, Reading, Massachusetts.

Simon, H. A.: 1966, 'A Note on Almost-Everywhere Definability' (abstract), *Journal of Symbolic Logic* 31, 705–706.

Simon, H. A.: 1970, 'The Axiomatization of Physical Theories', *Philosophy of Science* **37**, 16–26.

Simon, H. A. and Groen, G. J.: 1973, 'Ramsey Eliminability and The Testability of Scientific Theories', *British Journal of The Philosophy of Science* **24**, 367–380.

Sneed, J. D.: 1971, *The Logical Structure of Mathematical Physics*, D. Reidel Publishing Company, Dordrecht, Holland.

Suppes, P.: 1959, 'Measurement, Empirical Meaningfulness, and Three-Valued Logic', in C. W. Churchman and P. Ratoosh (eds.), *Measurement: Definitions and Theories*, John Wiley & Sons, New York, pp. 129–143.

Tuomela, R.: 1968, 'Identifiability and Definability of Theoretical Concepts,' *Ajatus* **30**, 195–220.

Tuomela, R.: 1973, *Theoretical Concepts*, Springer-Verlag, Vienna.

V. N. SADOVSKY AND V. A. SMIRNOV

DEFINABILITY AND IDENTIFIABILITY: CERTAIN PROBLEMS AND HYPOTHESES

Recently there has been definite progress in the development of definability and identifiability theory – both in logical and methodological spheres (Przełęcki, 1969; Simon, 1970; Hintikka, 1972; Tuomela, 1973; Rantala, 1975). On the other hand, up to the present time we still do not have at our disposal a completely rigorous theory of definability and identifiability – there is a variety of quite different kinds of approach to the construction of such a theory (which, incidentally, is quite natural), and moreover, the basic notions (and first of all the meaning of the concepts 'identifiability' and some others) are interpreted by different authors in essentially different ways

We believe that one of the problems facing the present Symposium – along with the discussion of specific logical and methodological aspects of definability and identifiability – is the elucidation of the basic notions of such a theory.

1. BASIC NOTIONS. THE CLASSICAL CONCEPT OF DEFINABILITY. FORMULATION OF THE PROBLEMS

The appearance of the modern theory of definitions and definability can be traced to the fundamental works of Padoa (1900), Tarski (1934), and Beth (1953). Considerable progress was made when it became possible to prove the equivalence of the syntactic and semantic concepts of definability, to compare the definition theory with the theory of deduction and to learn how to reduce the question of definition to that of deduction.

Let us introduce the necessary syntactic and semantic notions.

Let L be *the first-order language with identity, A* – its *vocabulary*, i.e. the set of descriptive symbols. We shall call *interpretation J* of the language L into a set X a function establishing correspondence between each individual constant from A and an individual from X, a

Butts and Hintikka (eds.), Basic Problems in Methodology and Linguistics, 63–80.
Copyright © 1977 by D. Reidel Publishing Company, Dordrecht-Holland. All Rights Reserved.

k-place predicate constant P and an element from 2^{X^k}, etc. By *possible realization of the language L, restricted by a vocabulary A'* we mean a triple $\langle X, J, A' \rangle$ where X is a non-empty set, J – interpretation of L into X and $A' \subseteq A$.

The concept of *truth in a possible realization* $M = \langle X, J, A \rangle$ is defined in the usual way. The expression $M \vDash B$ reads: B is true in M, M is *a model* of B. $M \vDash \Sigma$ means that every sentence from Σ is true in M.

The term *theory* will stand for a set of sentences closed under deduction. Σ is a theory iff $\forall B(\Sigma \vdash B \supset B \in \Sigma)$. Each set of sentences Σ is set in correspondence with a theory $\mathrm{Cn}(\Sigma) = \{B \mid \Sigma \vdash B\}$. Note that the models of Σ and $\mathrm{Cn}(\Sigma)$ are the same. $\mathrm{Mod}(\Sigma)$ is a set of all models of Σ. Let $\mathfrak{M}$ be a class of possible realizations, then $\mathrm{Tr}(\mathfrak{M})$ is a set of sentences that are true in each realization from $\mathfrak{M}$. Each theory Σ may be regarded as a set of all truth sentences in a certain class of possible realizations.

The notions of *identity* ($=$), *isomorphism* ($\simeq$) and *elementary equivalence* ($\equiv$) are introduced in the usual manner. A possible realization $\langle X, J, A \rangle$ will be called the *reduct* of realization $\langle X, J, A' \rangle$ if $A \subset A'$. If M is a reduct of M', M' is called an *expansion* of M. The expansion M is also denoted by (M, R) where R is the new relation. The reduct obtained from M' by elimination of relation R will also be denoted by $M'(\hat{R})$.

In the present paper, when discussing the problems of definability and identifiability, we shall use everywhere the notion of a predicate term P^l. The expressions Px and $\forall x Px$ will stand for the abbreviation of $P(x_1, ..., x_l)$ and $\forall x_1 \cdots \forall x_l P(x_1, ..., x_l)$, respectively.

The classical theory of definability will be meant as the theory of definability constructed in the works of Padoa, Tarski and Beth.

The classical theory of definability considers the definability of a term P in terms of a set of sentences Σ. This theory is based on the notions of *explicit definability, semantic definability* and *implicit definability*, which are listed in Table I.

Between these notions there exist, as is well known, the following relations:

THE PADOA THEOREM.   *If P is explicitly definable in terms of Σ, P is semantically definable in terms of Σ.*

TABLE I

Explicit definability $\Sigma \vdash \forall x (Px \sim \alpha)$	
Implicit definability $\Sigma, \Sigma' \vdash \forall x (Px \sim P'x),$ $\Sigma' = F_{P'}^{P} \Sigma$	Semantic definability $\forall J' \forall J'' (\langle X, J', A \cup \{Px\}\rangle \vDash \Sigma$ $\& \langle X, J'', A \cup \{Px\}\rangle \vDash \Sigma$ $\& \langle X, J', A \rangle = \langle X, J'', A \rangle$ $\supset \langle X, J', A \cup \{Px\}\rangle$ $= \langle X, J'', A \cup \{Px\}\rangle)$

THE BETH THEOREM. *If P is semantically definable in terms of Σ, P is explicitly definable in terms of Σ.*

The equivalence of implicit definability and semantic definability is sufficiently obvious; the implicit definability is essentially a reformulation of the requirements of semantic definability in syntactic terms. Therefore the Beth theorem is usually formulated in terms of explicit and implicit definability:

THE BETH THEOREM. *If P is implicitly definable in terms of Σ, P is explicitly definable in terms of Σ.*

Thus, all the three concepts – explicit, implicit and semantic definability – are equivalent.

It is necessary to point out that the terms explicit, implicit, and semantic definability mean the various *modes of definability:* in our case, the various *modes of complete definability*, since the definitions of these notions, given in Table I, satisfy the requirements of complete definition (non-creativity and eliminability). Naturally, a question arises concerning the possibility of supplementing these modes by still another one, which may be called the *explicit semantic definability*.

The main argument supporting the formulation of this question is that in the classical theory of definability, as was already noted, the notion of semantic definability is essentially that of implicit semantic definability, and therefore the tendency to find the semantic counterpart of explicit syntactic definability appears natural. In the present paper, we are mainly concerned with obtaining a positive solution to this problem; the result of this is the formulation of four different modes of complete definability, which are listed in Table II in the terminology introduced above:

TABLE II

Explicit syntactic definability	Explicit semantic definability
Implicit syntactic definability	Implicit semantic definability

As is well known, the classical theory of complete definability of Padoa–Tarski–Beth was generalized in the works of Svenonius (1959), Chang (1964), Makkai (1964), and Kueker (1970). In these works the authors introduced *the weaker* (as compared with complete explicit definability) kinds of syntactic definability (piecewise, relative and restricted) and demonstrated the equivalence between these kinds of definability and the corresponding implicit semantic definability requirements (for the summary of these results, see, for instance, Rantala, 1975). We shall call all kinds of definability that are weaker than complete definability '*incomplete definability*'. With respect to those we shall formulate two more questions (in addition to the previously posed ones), which will be in the focus of our attention as far as the present paper is concerned:

Is it possible to formulate the explicit semantic counterparts for the notions of explicit syntactic piecewise, relative and restricted definability?

Is it possible to construct for the notions of the explicit syntactic piecewise, relative and restricted definability the respective counterparts in terms of implicit syntactic definability?

Below we shall present the arguments in favor of the positive answer (at least partially) to the above two questions. Such an answer provides an opportunity of formulating generalizations of the notion of definability; it proves possible to construct on its basis certain important relations between various modes and kinds of definability which promotes clarification of the foundations of the theory of definability and identifiability.

In what follows we shall use certain terminological abbreviations. Instead of the expression 'complete definability' we shall write simply 'definability' (in those cases, of course, that would not lead to confusion). The weaker kinds of definability will always be denoted in full (piecewise definability, etc.). According to tradition, the kind of definability that in our terminology is called 'explicit syntactic complete definability' is designated as 'explicit definability'; in our paper we shall also use this abbreviation. Since in this case the term 'explicit definability' means a definite kind of definability rather than a mode of definability (explicit syntactic or explicit semantic), there is no danger of confusion of different terminological meaning in the corresponding contexts.

2. DEFINABILITY IN MODELS. PIECEWISE DEFINABILITY

In order to find the answer to the first two questions formulated in the preceding section, let us introduce the notion of *definability of a term in a possible realization*:

P is definable in a possible realization M iff there is such a sentence α with exactly l free variables, formulated in terms different from P, that $\forall x(Px \sim \alpha)$ is a truth in M, or, in symbolic form:

$$P \text{ is } \textit{definable in } M \rightleftharpoons (M \vDash \forall x(Px \sim \alpha)).$$

Now we are in a position to express the concept of explicit definability in terms of the model theory (in semantic terms).

A term P is *explicitly semantically definable in terms of a set of sentences* Σ iff there is such a formula α that defines P in each model of Σ:

$$\exists \alpha \underset{M \in \mathrm{Mod}(\Sigma)}{\forall M} \ (M \vDash \forall x(Px \sim \alpha)).$$

The definability in each model of Σ does not yet mean the explicit semantic definability in Σ; in order for such definability to take place the term P should be defined in all models by the same formula α.

It is obvious that the new notion of definability is equivalent to three other basic notions of the classical theory of definability. In the frame of the theory of complete definability this notion plays no important role, but their analogues for generalized concepts of definability, as we see it, are undoubtedly interesting.

The reformulation of explicit syntactic definability in terms of the model theory opens up a possibility for generalization of the notion of definability.

Let K be a set of formulas with l free variables in each. We shall introduce the concepts of the piecewise, finite piecewise and n-piecewise (for a fixed n) definability based on the notion of explicit semantic definability. The concept of 1-piecewise definability coincides with that of explicit complete definability. The types of piecewise definability that are introduced – in explicit syntactic and semantic formulations – are given in Table III.

TABLE III

Explicit syntactic formulations	Explicit semantic formulations
Explicit complete (1-piecewise) definability	
$\exists\alpha(\Sigma \vdash \forall x(Px \sim \alpha))$	$\displaystyle\exists\alpha \mathop{\forall M}_{M\in\mathrm{Mod}(\Sigma)} (M \vDash \forall x(Px \sim \alpha))$
	$\displaystyle\mathop{\exists K}_{|K|\leqslant 1} \mathop{\forall M}_{M\in\mathrm{Mod}(\Sigma)} \mathop{\exists\alpha}_{\alpha\in K} (M \vDash \forall x(Px \sim \alpha))$
n-piecewise definability	
$\exists\alpha_1 \ldots \exists\alpha_n(\Sigma \vdash \mathop{\bigvee}_{1\leqslant i\leqslant n} \forall x(Px \sim \alpha_i))$	$\displaystyle\mathop{\exists K}_{|K|\leqslant n} \mathop{\forall M}_{M\in\mathrm{Mod}(\Sigma)} \mathop{\exists\alpha}_{\alpha\in K} (M \vDash \forall x(Px \sim \alpha))$
Finite-piecewise definability	
$\displaystyle\mathop{\exists n}_{n\in N} \exists\alpha_1 \ldots \exists\alpha_n(\Sigma \vdash \mathop{\bigvee}_{1\leqslant i\leqslant n} \forall x(Px \sim \alpha_i))$	$\displaystyle\mathop{\exists n}_{n\in N} \mathop{\exists K}_{|K|\leqslant n} \mathop{\forall M}_{M\in\mathrm{Mod}(\Sigma)} \mathop{\exists\alpha}_{\alpha\in K} (M \vDash \forall x(Px \sim \alpha))$
Piecewise definability	
	$\displaystyle\mathop{\forall M}_{M\in\mathrm{Mod}(\Sigma)} \exists\alpha(M \vDash \forall x(Px \sim \alpha))$

It is clear that if $m < n$, from m-piecewise definability follows n-piecewise definability; if a term is n-piecewise definable it is also finite-piecewise definable.

From finite-piecewise definability follows piecewise definability. The Svenonius theorem (Svenonius, 1959) leads to the conclusion that from piecewise definability follows finite-piecewise definability. Thus the above two notions are equivalent.

The formulation of the syntactic notion of the piecewise definability in explicit semantic terms makes it possible to obtain a relatively simple proof of the following theorem:

THEOREM 1. *If a term P is piecewise definable in a complete theory Σ, it is explicitly completely definable in Σ.*

(The proofs of the numbered theorems are given in Sadovsky and Smirnov, 1977.)

This theorem provides still another way of establishing the incompleteness of a theory: if a certain term of a theory Σ is piecewise definable but not explicitly completely definable in it, this theory is incomplete.

3. PARAMETRIC DEFINABILITY

In this section we shall present semantic formulations of the syntactic concept of restricted definability (the Chang–Makkai definability). Let us first introduce the notion of *parametric definability in a model:* a formula α with l free variables $x_1, ..., x_l$ and m parameters $y_1, ..., y_m$ parametrically defines a predicate constant P^l in a model M iff

$$\underset{a \in M}{\exists a}\, (M \vDash \forall x(Px \sim \alpha x a)).$$

In addition to abbreviations previously introduced we shall write $\alpha x a$ instead of $\alpha(x_1, ..., x_l, a_1, a_2, ..., a_m)$.

Parametric definability of the term P in a model M is identical with its explicit definability in the expansion $(M, a)_{a \in X}$ where X is the basic set of M. Thus, the various kinds of parametric definability are based on the usual notion of explicit definability, but this notion is applied in the expansions by individuals of models Σ instead of the models of system Σ themselves.

Making use of the notion of parametric definability we can – by analogy with how it was done in the analysis of the piecewise definability – introduce various forms of explicit parametric (restricted) definability that are listed in the Table IV.

TABLE IV

Explicit syntactic formulations	Explicit semantic formulations
1-parametric definability	
$\exists\alpha(\Sigma \vdash \exists y \forall x(Px \sim \alpha xy))$	$\underset{M\in\mathrm{Mod}(\Sigma)}{\exists\alpha} \ \forall M \ \underset{a\in M}{\exists a} \ (M\vDash\forall x(Px \sim \alpha xa))$
	$\underset{|K|\leqslant 1}{\exists K} \ \underset{M\in\mathrm{Mod}(\Sigma)}{\forall M} \ \underset{\alpha\in K}{\exists\alpha} \ \underset{a\in M}{\exists a} \ (M\vDash\forall x(Px \sim \alpha xa))$
n-parametric definability	
$\exists\alpha_1\ldots\exists\alpha_n\Big(\Sigma \vdash \underset{1\leqslant i\leqslant n}{\bigvee} \exists y \forall x(Px \sim \alpha_i xy)\Big)$	$\underset{|K|\leqslant n}{\exists K} \ \underset{M\in\mathrm{Mod}(\Sigma)}{\forall M} \ \underset{\alpha\in K}{\exists\alpha} \ \underset{a\in M}{\exists a} \ (M\vDash\forall x(Px \sim \alpha xa))$
Finite parametric definability	
$\underset{n\in N}{\exists n} \ \exists\alpha_1\ldots\exists\alpha_n\Big(\Sigma$	$\underset{n\in N}{\exists n} \ \underset{|K|\leqslant n}{\exists K} \ \underset{M\in\mathrm{Mod}(\Sigma)}{\forall M} \ \underset{\alpha\in K}{\exists\alpha} \ \underset{a\in M}{\exists a}(M$
$\vdash \underset{1\leqslant i\leqslant n}{\bigvee} \exists y \forall x(Px \sim \alpha_i xy)\Big)$	$\vDash\forall x(Px \sim \alpha xa))$
Parametric definability	
	$\underset{M\in\mathrm{Mod}(\Sigma)}{\forall M} \ \exists\alpha \ \underset{a\in M}{\exists a} \ (M\vDash\forall x(Px \sim \alpha xa))$
	$\underset{M\in\mathrm{Mod}(\Sigma)}{\exists K} \ \forall M \ \underset{\alpha\in K}{\exists\alpha} \ \underset{a\in M}{\exists a} \ (M\vDash\forall x(Px \sim \alpha xa))$

Note that in Table IV in the section 'explicit syntactic formulations' we present Chang's and Keisler's notions of definability explicitly up to parameters (1-parametric definability) and definability explicitly up to parameters and disjunction (n-parametric and finite parametric definability) (Chang and Keisler, 1973, p. 250).

Finite parametric definability is the Chang–Makkai, or the restricted definability. We don't have an answer to the question whether finite parametric definability is equivalent to parametric definability. (In discussion at the symposium J. Hintikka said that V. Rantala has come up with a positive answer to this question.)

4. Conditional and conditional-parametric definability

In order to construct the explicit semantic formulations of the relative definability (the Kueker definability) we shall introduce the notions of the conditional and conditional parametric definability.

As is well known, *the conditional definition* is a definition such as

$$\forall x(Sx \supset (Px \sim \alpha x)).$$

A conditional definition specifies the value of a predicate P only over the range S. Outside of this range the predicate P is not specified. A convenient technique for formulation of this kind of definitions is that of restricted quantifiers

$$\forall_{Sx} x(Px \sim \alpha x).$$

In mathematics and the empirical sciences we quite frequently come across definitions with a restricted quantifier. Often in the quantifier-free form they are written as

$$Ax \& Px \sim Ax \& \alpha x,$$

where Ax specifies the range of specification of the predicate P and αx is a formula with a free variable x. According to Gorsky, 1974, these definitions form a special class, however they are equivalent to

$$Ax \supset (Px \sim \alpha x).$$

Note that the traditional definition through the genus and differentia specifica can be introduced by means of an indication of the respective range of the predicate (genus) and the defining condition, in other words, as a conditional definition.

We shall say that a predicate term P is *syntactically conditionally definable* in terms of a set of sentences Σ iff

$$\exists \alpha \left(\Sigma \vdash \forall_{Sx} x(Px \sim \alpha x) \right).$$

Definitions introduced through restricted quantifiers, i.e., the conditional definitions, are not only widely used in the empirical sciences but appear naturally in the transition from many-sorted to one-sorted theories. Suppose that a term P is introduced in a many-sorted theory Σ through an explicit definition $\forall \eta(P\eta \sim \alpha\eta)$. In the standard approach to a one-sorted theory this definition is transformed into $\forall x(Px \sim \alpha x)$.
Sx

Unifying the conditional and the parametric definabilities (with the requirement of the non-emptiness of the defining condition being satisfied) we arrive at the *conditional-parametric definability*.

Introduce for convenience a new quantifier WxA:
Bx

$$WxA \rightleftharpoons \exists xBx \,\&\, \forall x(Bx \supset A).$$
Bx

From the definition of this quantifier it follows that

$$WxA \supset \exists xA$$
Bx

$$\exists xBx \,\&\, \forall xA \supset WxA$$
Bx

$$WxA \supset \exists x(Ax \,\&\, Bx)$$
Bx

We shall say that a term P is *conditional-parametric definable in M* iff there are such formulas α and S, formulated in terms different from P, that

$$Wa(M \vDash \forall x(Px \sim \alpha xa)).$$
$M \vDash Sa$

Conditional-parametric definability of the term P in model M is identical with its explicit definability in each expansion $(M, a_1, ..., a_k)$ where $a_1, ..., a_k$ is the sequence of individuals from the basic set X, which satisfies the condition S (of non-emptiness). Thus, the notion of finite identifiability is based on the ordinary concept of definability in all expansions of M by k-tuple individuals, which satisfy a certain condition S.

The Table V lists various kinds of conditional-parametric definability.

TABLE V

Explicit syntactic formulations	Explicit semantic formulations
Conditional-1-parametric definability	
$\exists S \exists \alpha (\Sigma \vdash \exists y Sy \ \& \ \forall y (Sy \supset \forall x (Px \sim \alpha xy)))$	$\exists S \exists \alpha \ \underset{M \in \mathrm{Mod}(\Sigma)}{\forall M} \ \underset{M \models Sa}{Wa} (M \models \forall x (Px \sim \alpha xa))$
Conditional-n-parametric definability	
$\exists S \exists \alpha_1 \ldots \exists \alpha_n \Big(\Sigma \vdash \exists y Sy$ $\& \underset{1 \leqslant i \leqslant n}{\bigvee} \forall y (Sy \supset \forall x (Px \sim \alpha_i xy)) \Big)$	$\exists S \ \underset{\lvert K \rvert \leqslant n}{\exists K} \ \underset{M \in \mathrm{Mod}(\Sigma)}{\forall M} \ \underset{M \models Sa}{Wa} (M$ $\models \forall x (Px \sim \alpha xa))$
Conditional finite parametric definability	
$\exists S \exists n \exists \alpha_1 \ldots \exists \alpha_n \Big(\Sigma \vdash \exists y Sy$ $\& \underset{1 \leqslant i \leqslant n}{\bigvee} \forall y (Sy \supset \forall x (Px \sim \alpha_i xy)) \Big)$	$\exists S \ \underset{n \in N}{\exists n} \ \underset{\lvert K \rvert \leqslant n}{\exists K} \ \underset{M \in \mathrm{Mod}(\Sigma)}{\forall M} \ \underset{\alpha \in K}{\exists \alpha} \ \underset{M \models Sa}{Wa} (M$ $\models \forall x (Px \sim \alpha xa))$
Conditional-parametric definability	
	$\exists S \ \underset{M \in \mathrm{Mod}(\Sigma)}{\forall M} \ \underset{M \models Sa}{\exists \alpha \ Wa} (M \models \forall x (Px \sim \alpha xa))$

The conditional-1-parametric definability, as is well known, is equivalent to the explicit complete definability.

Assuming that there is a non-empty class of conditions δ, and imposing restrictions on its cardinality. we obtain a new type of definability

$$\underset{\lvert \delta \rvert \leqslant m}{\exists \delta} \ \underset{\lvert K \rvert \leqslant n}{\exists K} \ \underset{M \in \mathrm{Mod}(\Sigma)}{\forall M} \ \underset{S \in \delta}{\exists S} \ \underset{\alpha \in K}{\exists \alpha} \ \underset{M \models Sa}{Wa} (M \models \forall x (Px \sim \alpha xa))$$

(see also Chang and Keisler, 1973, p. 255).

If $m = 1$ and $n = 1$ we have explicit complete definability; if $n = 1$

this type of definability is equivalent to piecewise definability and if $m = 1$ it is identical with conditional-parametric definability.

Thus, we have obtained a positive answer to the first two questions formulated in Section 1. For all types of explicit syntactic definability (complete, piecewise, relative, restricted, etc.) the respective explicit counterparts are constructed, that are formulated in semantic terms. These semantic formulations provide a convenient basis for classification of various types of definability.

5. IMPLICIT SYNTACTIC DEFINABILITY

Let us consider our third problem (see p. 66). As is well known, the proof of equivalence of the explicit and implicit definability makes it possible to reduce the issue of definability to that of deduction. Of course, it would be very desirable to solve a similar problem for the generalized forms of definability. We shall answer this question in the case of two types of definability – the conditional and the piecewise.

A predicate term P is *implicitly conditionally definable* in terms of a set of sentences Σ iff

$$\Sigma, \Sigma' \vdash \underset{Sx}{\forall} x(Px \sim P'x)$$

The following theorem is valid:

THEOREM 2. *A term P is explicitly conditionally definable in terms of a set of sentences Σ iff it is implicitly conditionally definable in Σ.*

A predicate term P is *implicitly n-piecewise definable* in terms of a set of sentences Σ iff

$$\Sigma^0, \Sigma^1, ..., \Sigma^n \vdash \bigvee_{1 \leqslant j \leqslant n} \bigvee_{1 \leqslant i \leqslant n} \forall x(P^i x \sim P^j x)$$

where Σ^i is the result of substitution of a term P^i in place of a term P in Σ, $\Sigma^0 = \Sigma$ and $P^0 = P$.

For $n = 2$ the implicit piecewise definability takes the form

$$\Sigma, \Sigma', \Sigma'' \vdash \forall x (Px \sim P'x) \vee \forall x (Px \sim P''x) \vee \forall x (P'x \sim P''x).$$

There is a following theorem:

THEOREM 3. *A term P is explicitly n-piecewise definable in terms of a set of sentences Σ iff P is implicitly n-piecewise definable in Σ.*

Is it possible to formulate the notions of implicit syntactic definability corresponding to the Kueker (conditional-parametric) or the Chang–Makkai (parametric) types? This question we shall leave open for discussion.

6.   IMPLICIT SEMANTIC DEFINABILITY. DEFINABILITY AND IDENTIFIABILITY

The term 'implicit semantic definability' will mean – in accordance with the adopted terminology – various types of semantic (in the traditional sense) definability (identifiability). The fundamental results in this field are obtained by Svenonius (1959), Kueker (1970), Chang (1964), and Makkai (1964).

Let $M = \langle X, J, A \cup \{P\} \rangle$ be a *model* of Σ, and $M(\hat{P}) = M^* = \langle X, J, A \rangle$ is a *reduct* of M.

We shall form now a *class of relations* with which $J(P)$ can be identified:

$$U(M(\hat{P})) = U(\langle X, J, A \rangle)$$
$$= \{R \mid \exists J'(R = J'(P, X)) \,\&\, \langle X, J', A \cup \{P\} \rangle$$
$$\in \text{Mod}\,(\Sigma) \,\&\, \langle X, J', A \rangle = \langle X, J, A \rangle \}.$$

If for each model M of the set Σ the cardinality of the class $U(M(\hat{P}))$ is less or equal to n where $n \geq 1$, P is called *n-foldly identifiable*.

P is *finitely identifiable* in Σ if there is such an n that P is n-foldly identifiable in Σ.

An important case of 1-*identifiability* coincides with the notion of the explicit definability of P.

P^l is *restrictedly identifiable* in Σ iff for any infinite model M

$$|U(M(\hat{P}))| < 2^{\aleph^l}.$$

Each of the above types of identifiability, as is well known, corresponds to a certain form of definability formulated in syntactic terms (Table VI).

TABLE VI

Explicit syntactic formulations	Implicit semantic formulations (in terms of identifiability)
Explicit complete definability $\exists\alpha(\Sigma \vdash \forall x(Px \sim \alpha))$	1-identifiability $\|U(M(\hat{R}))\| = 1$
$\exists S \exists \alpha_1 \ldots \exists \alpha_n \Big(\Sigma \vdash \exists y Sy$ $\& \, \forall y \Big(Sy \supset \bigvee_{1 \leq i \leq n} \forall x(Px \sim \alpha_i xy) \Big) \Big)$ Kueker	n-identifiability $\|U(M(\hat{R}))\| \leq n$
$\exists \alpha_1 \cdots \exists \alpha_n \Big(\Sigma \vdash \bigvee_{1 \leq i \leq n} \exists y \forall x(Px \sim \alpha_i xy) \Big)$ Chang–Makkai	Restricted identifiability $\|U(M(\hat{R}))\| < 2^{\aleph^l}$

According to Svenonius (1959), a term P is *finitely-piecewise* definable in Σ iff two isomorphic models having the same reduct are identical.

As can be seen from the table presenting the summary of various types and forms of definability and identifiability (see Table VII), in the column of implicit semantic definability there are many empty spaces. The situation is best in the Kueker's section of the table.

We believe that the first question that has to be investigated is that of the implicit semantic analogue of the n-piecewise definability.

Leaving this issue open for discussion, let us note that from the n-piecewise definability follows n-identifiability. Indeed, every reduct

TABLE VII

Summary of various types of definability and identifiability

I – Explicit semantic formulations
II – Explicit syntactic formulations
III – Implicit semantic formulations
IV – Implicit syntactic formulations

	I	II	III	IV
1-piecewise = explicit complete	+	+	+	+
n-piecewise	+	+		+
Finite-piecewise = piecewise	+	+	+	+
Conditional-1-parametric = explicit complete	+	+	+	+
Conditional-n-parametric	+	+	+	
Conditional finite parametric	+	+	+	
Conditional-parametric	+			
1-parametric	+	+		
n-parametric	+	+		
Finite parametric	+	+	+	
Parametric	+			

of a certain model of Σ has not more than one expansion from the class $\{M \in \mathrm{Mod}\,(\Sigma) \mid M \vDash \forall x (Px \sim \alpha_i x)\}$ where $\alpha_i x \in K$, since if $M_1(\hat{P}) = M_2(\hat{P})$ and $\forall x (Px \sim \alpha_i x)$ is truth in M_1 and M_2, $M_1 = M_2$. Therefore if a term P is n-piecewise definable in Σ, P is n-foldly identifiable in Σ. In particular, from the Svenonius type definability follows the finite identifiability.

7. APPLICATION OF THE CONCEPT OF DEFINABILITY TO THE THEORY OF DEFINITION

The theory of definability deals directly not with the introduction of new terms in a theory but with establishing a certain type of interrelation between the terms already existing in the theory. However, the theory of definability, as we see it, should form the basis of constructing the theory of definition proper.

We shall say that a set of sentences Σ *plays the role of a definition* in a theory T of a new l-place term P (P is not in the vocabulary of the theory) iff there is a formula α formulated in terms of the theory T with l different free variables such that $Cn(T \cup \Sigma)$ is equivalent to

$$Cn(T \cup \{\forall x(Px \sim \alpha)\}).$$

There is a following theorem.

THEOREM 4. *A set of sentences Σ plays the role of a definition of a term P in a theory T iff*

(1)   *P is explicit completely definable in $T \cup \Sigma$,*
(2)   *$T \cup \Sigma$ is a conservative expansion of T.*

The point (1) is a condition of *eliminability*, (2) – that of *non-creativity*.

In the formulation of the theorem it is assumed that Σ contains one additional symbol P. In the general case (1) requires that each new term is definable in $T \cup \Sigma$ in terms of T.

The stated theorem and the theorem of equivalence between the explicit and the implicit definability make it possible to reduce the question of whether Σ plays the role of a definition in T to the question of deduction. First the inference $\Sigma, \Sigma' \vdash \forall x(Px \sim P'x)$ is sought. If this inference exists, from the theorem of the equivalence between the explicit and implicit definability we find the defining formula α.

The question of non-creativity is reduced to that of provability of each formula of Σ in $T \cup \{\forall x(Px \sim \alpha)\}$.

Certain sets of sentences play the role of a definition for arbitrary theories. Such are, for example, explicit definitions. Other sets of sentences play the role of a definition under certain conditions imposed on the theory.

Thus, a system of *bilateral reduction sentences*

$$\begin{cases} \forall x(A_1 x \supset (Px \sim \alpha_1)) \\ \quad \vdots \\ \forall x(A_n x \supset (Px \sim \alpha_n)) \end{cases}$$

plays the role of a definition in a theory in which it is provable that the testability conditions $A_1 x, \ldots, A_n x$ are (1) exhaustive ($T \vdash$

$\forall x (A_1 x \lor \cdots \lor A_n x)$ and (2) mutually incompatible ($\forall i \forall j (i \neq j \supset \exists x (A_i x \,\&\, A_j x))$). The term P occurs neither in A_i nor in α_i.

The first condition is necessary for the proof of definability of P (eliminability of P), the second – in order to guarantee that the extension through adding the system of bilateral reduction definitions would be non-creative (Smirnov, 1974).

Definitions by means of a system of bilateral reduction sentences are widely used in mathematics. These definitions are more frequently called *definitions by cases*.

For example, for defining functions they are written in the form

$$f(x) = \begin{cases} \varepsilon(x), & \text{if } x > 0 \\ \varphi(x), & \text{if } x = 0 \\ \psi(x), & \text{if } x < 0 \end{cases}$$

It is necessary that the conditions in definitions by cases be mutually incompatible and exhaustive.

Academy of the Sciences of the USSR

BIBLIOGRAPHY

Beth, E. W.: 1953, 'On Padoa's Method in the Theory of Definition', *Indagationes Mathematicae* **15**, 330–339.
Chang, C. C.: 1964, 'Some New Results in Definability', *Bulletin of the American Mathematical Society* **70**, 808–813.
Chang, C. C. and Keisler, H. J.: 1973, *Model Theory*, North-Holland, Amsterdam.
Gorsky, D. P.: 1974, *Definition*, Publishing House 'Mysl', Moscow (in Russian).
Hintikka, J.: 1972, 'Constituents and Finite Identifiability', *Journal of Philosophical Logic* **1**, 45–52.
Kalman, R. E.: 1960, 'On the General Theory of Control Systems', in *First IFAC Congress*, Moscow, pp. 481–492.
Kueker, D. W.: 1970, 'Generalized Interpolation and Definability', *Annals of Mathematical Logic* **1**, 423–468.
Makkai, M.: 1964, 'A Generalization of a Theorem of E. W. Beth', *Acta Math. Acad. Sci. Hungar.* **15**, 227–236.
Padoa, A.: 1900. 'Essai d'une théorie algebrique des nombres entiers, précedé d'une introduction logique à une théorie deductive quelconque, secs. 8–18', *Bibliothèque du Congrès International de Philosophie* **3**, Paris, p. 309.
Przełęcki, M.: 1969, *The Logic of Empirical Theories*, Routledge and Kegan Paul, London.

Rantala, V.: 1975, 'Definability Problems in the Methodology of Science', in *Conference for Formal Methods in the Methodology of Empirical Sciences, Warsaw, 1974*, Warsaw.

Sadovsky, V. N. and Smirnov, V. A.: 1977, 'Complete and Incomplete Definability in First-order Theories', in *The Methods of Logical Analysis of Scientific Language*, Publishing House 'Nauka', Moscow (in Russian).

Simon, H. A.: 1970, 'The Axiomatization of Physical Theories', *Philosophy of Science* **37**, 16–26.

Smirnov, V. A.: 1974, 'On the Question of Definability of Predicates Introduced by Bilateral Reduction Sentences', in *Philosophy in the Modern World. Philosophy and Logic*, Publishing House 'Nauka', Moscow, pp. 165–167 (in Russian).

Svenonius, L.: 1959, 'A Theorem about Permutation in Models', *Theoria* **25**, 173–178.

Tarski, A.: 1934, 'Some Methodological Investigations on the Definability of Concepts', in A. Tarski, *Logic, Semantics, Metamathematics*, Clarendon Press, Oxford, 1956, pp. 296–319.

Tuomela, R.: 1973, *Theoretical Concepts*, Springer Verlag, Wien-New York.

MARIAN PRZEŁĘCKI

ON IDENTIFIABILITY IN EXTENDED DOMAINS

For the sake of the present discussion, by definability I shall understand what is usually called the explicit (or syntactic) definability; by
identifiability – the implicit (or semantic) definability. I shall restrict
myself to considering empirical theories which can be formalized in
first-order logic. As it is known, with regard to such theories these two
concepts coincide. A term t is definable in a theory T iff it is
identifiable in it. That is to say, a definition of t is a theorem of T iff
the interpretation of the remaining terms in any model of T fixes
uniquely the interpretation of t. That is why all cases of non-
identifiability considered thus far are always connected with some kind
of non-definability. A concept is not identifiable only if it is not
definable; e.g. if it is governed by some postulate weaker than an
explicit definition (a conditional, or piecewise, or partial definition – to
mention the typical cases). In what follows, I wish to call attention to
some other kind of non-identifiability and its characteristic sources.
Non-identifiability of this kind is not connected with non-definability.
It appears as a result of some looser conception of a theory's interpretation. On the standard approach, any extension of a given language
has as its semantic counterpart a suitable expansion of the language's
structures. But there seem to be cases where an extension of language
is connected not only with expanding its structures, but also with
extending their domains. This seems to be the case when an observational language is being extended by introducing into it certain types of
theoretical terms. In such cases the correspondence between definability and identifiability mentioned above does not apply any longer.
Being definable in a given theory, a theoretical concept is not identifiable in the theory's models – if these are structures with extended
domains.

To describe the situation more closely, we shall avail ourselves of the
usual model theoretic terminology. Let L_o be a first-order language

Butts and Hintikka (eds.), Basic Problems in Methodology and Linguistics, 81–89.

with $o_1, ..., o_n$ as its only descriptive constants, and L an extension of L_o containing t as an additional descriptive constant. Structures for L_o will be symbolized by $\mathfrak{M}_o$ (or $\mathfrak{N}_o$), those for L by $\mathfrak{M}$ (or $\mathfrak{N}$). The domain of a structure $\mathfrak{M}$ will be denoted by $|\mathfrak{M}|$, the interpretation in $\mathfrak{M}$ of a constant c by $c^{\mathfrak{M}}$. By $\mathfrak{M}|_o$ we shall denote the restriction (the reduct) of structure $\mathfrak{M}$ to language L_o; $\mathfrak{M}$ is then called an expansion of $\mathfrak{M}|_o$. If T is a set of sentences of L, the class of its models will be symbolized by Mod (T).

Now, according to the standard approach, the extending of language L_o to language L is thought of as accompanied by the expanding of structures $\mathfrak{M}_o$ to structures $\mathfrak{M}$. Hence, in examining the relationship between (syntactic) definability and (semantic) identifiability, what we take into account is a class of structures, $M_{\mathfrak{M}_o}$, defined, for any $\mathfrak{M}_o$, as follows:

(I) $M_{\mathfrak{M}_o} = \{\mathfrak{M}: \mathfrak{M}|_o = \mathfrak{M}_o$ and $\mathfrak{M} \in$ Mod $(T)\}$.

$M_{\mathfrak{M}_o}$ comprises those expansions of $\mathfrak{M}_o$ which are models of T. Now, t is definable in T (in terms of $o_1, ..., o_n$) iff for any $\mathfrak{M}_o$, $M_{\mathfrak{M}_o}$ contains at most one structure. If T reduces to a definition of t (in terms of $o_1, ..., o_n$) then for any $\mathfrak{M}_o$, $M_{\mathfrak{M}_o}$ contains exactly one structure: for each $\mathfrak{M}_o$, there is exactly one interpretation of t. t is then said to be identifiable in T.

This approach is applicable to all situations in which the introduction of new terms does not involve any extension of the old universe of discourse. But, as far as empirical theories are concerned, not every procedure of introducing new terms seems to fall under such schema. There appear to be cases when the procedure is essentially connected with an enlargement of the theory's domain. A typical situation of this kind may be characterized as follows. Let L_o be an observation language and $o_1, ..., o_n$ observation predicates. Whatever these epithets may mean, we shall assume that any intended model $\mathfrak{M}_o$ for L_o consists of observables only: its domain $|\mathfrak{M}_o|$ is a set of observable things. Let t be a theoretical predicate (say 'electron') which is to refer to some unobservable, 'theoretical', entities. Introducing it into language L_o, we have to extend its intended models so as to include some unobservable objects into their domains; such structures only may be reckoned among the intended models for L.

In cases like this, it is the interpretation of t in the structures suitably extended that is of primary interest to us. What kind of extension is involved here? In model theory, $\mathfrak{N}_o$ is called an extension of $\mathfrak{M}_o$ iff $|\mathfrak{M}_o|$ is included in $|\mathfrak{N}_o|$ and $o_i^{\mathfrak{M}_o}$ is identical with $o_i^{\mathfrak{N}_o}$ restricted to $|\mathfrak{M}_o|$ (for $i = 1, ..., n$). So outside $|\mathfrak{M}_o|$, o_i may be interpreted in $\mathfrak{N}_o$ in any way whatsoever. Is this how observation predicates are actually interpreted within unobservable domains? As the concept of observation predicate is a notoriously ambiguous one, the answer clearly depends on how it is understood. Three main conceptions may be distinguished here. Within two of them, an observation predicate is conceived of as interpreted in a purely ostensive way. In consequence, its interpretation is assumed to be 'restricted' to observable things only. That 'restriction', however, is understood in two different ways.

(i) According to one explication, the interpretation of an observational predicate is assumed to be completely undetermined with regard to all unobservable objects: outside the observable domain the predicate may be interpreted in any way whatsoever. On this assumption, any extension $\mathfrak{N}_o$ of a structure $\mathfrak{M}_o$ will be taken into account in defining the intended models for L.

(ii) According to another explication, the interpretation of an observational predicate is assumed to be negatively determined with regard to all unobservable objects: the predicate is supposed to be true of observable things only, and so false of any unobservable entity. On this assumption, only those extensions $\mathfrak{N}_o$ of a structure $\mathfrak{M}_o$ will be taken into account in defining the intended models for L in which $o_i^{\mathfrak{N}_o} = o_i^{\mathfrak{M}_o}$ (for $i = 1, ..., n$).

(iii) The third conception is a more liberal one. It does not restrict observational predicates to terms interpreted in a purely ostensive way. An observational predicate is here assumed to be interpreted not only by some direct (non-verbal) methods, like ostension, but also in an indirect (verbal) way, viz. by means of some set of postulates. As far as unobservable domains are concerned, the indirect way of interpretation is supposed to be the only way available: the predicate may be interpreted directly within observable domains merely; outside them it may be interpreted only indirectly, through some set of postulates. Let set T_o of sentences of language L_o represent the set of postulates for the observation predicates $o_1, ..., o_n$. On the present assumption, only those extensions $\mathfrak{N}_o$ of a structure $\mathfrak{M}_o$ will be taken into account in defining the intended models for L which are models of postulates T_o: $\mathfrak{N}_o \in \mathrm{Mod}\,(T_o)$. To guarantee this, we shall require that the set T include the set T_o: $T_o \subseteq \mathrm{Cn}\,(T)$.[1]

Now, it is the last conception of observation predicates that I propose to adopt for the sake of the present discussion. It entails the following definition of class $N_{\mathfrak{M}_o}$, which is meant as a counterpart of class $M_{\mathfrak{M}_o}$, adapted to the case in question:

(II) $\qquad N_{\mathfrak{M}_o} = \{\mathfrak{M}\colon$ for some extension $\mathfrak{N}_o$ of $\mathfrak{M}_o$,

$$\mathfrak{M}|_o = \mathfrak{N}_o \text{ and } \mathfrak{M} \in \mathrm{Mod}\,(T)\}.$$

$N_{\mathfrak{M}_o}$ comprises those structures for L which are expansions of the extensions of $\mathfrak{M}_o$, and models of T.

The definition of $N_{\mathfrak{M}_o}$ reflects a fundamental feature of that conception of a theory's interpretation which underlies the present approach. The conception is known under the name of Semantical Empiricism. According to it, it is observation language only that can be interpreted directly, or, so to say, 'from without' – viz. by some ostensive, or operational, procedures. A theoretical language can be interpreted only indirectly and 'from within' – i.e. by means of postulates formulated in the language of a given theory. This is how all the theoretical terms are assumed to be interpreted. And, what seems even more important, this is how the theoretical universe is supposed to be determined. There is assumed to be no other way of determining the domain of theoretical entities than by stipulating that it be a domain which satisfies a given set of sentences of language L: the set T under our schema. This fact proves decisive in determining the character of the class $N_{\mathfrak{M}_o}$.

What is then the class $N_{\mathfrak{M}_o}$ like? How is the term t interpreted in its structures? There can be adduced a number of simple observations that provide a partial answer to this question. Let $\mathfrak{M}_o$ be a model of T_o (on our assumptions, only such structures may be regarded as intended models for L_o). Now, if only T is non-creative with respect to T_o, the class $N_{\mathfrak{M}_o}$ is bound to be non-empty. This is ensured by the following syntactic criterion of the non-emptiness of $N_{\mathfrak{M}_o}$:

$N_{\mathfrak{M}_o} \neq \varnothing$, for any $\mathfrak{M}_o \in \mathrm{Mod}\,(T_o)$, iff every purely universal L_o-consequence of T is a consequence of T_o.

At the same time, it is easily seen that if $N_{\mathfrak{M}_o}$ is non-empty, it always contains more than one structure – in fact, an infinite number of structures. And this is so independently of what the set T is like. So, even if t is definable in T, it always has an infinite number of interpretations in $N_{\mathfrak{M}_o}$, and in consequence, is not identifiable in T – in any plausible meaning (the only exception being the case when T implies the emptiness of t). Furthermore, even if $|\mathfrak{M}_o|$ consists of observable things, the set $|\mathfrak{M}| - |\mathfrak{M}_o|$ will, for some $\mathfrak{M} \in N_{\mathfrak{M}_o}$, consist of some abstract entities, e.g. numbers. Such entities then will make up the interpretation of t in some structures in $N_{\mathfrak{M}_o}$.[2]

Let us examine the interpretation of t in some detail (to simplify the notation, I shall assume that t is a monadic predicate). Now, whatever the set T may be like, the term t, as interpreted by the class $N_{\mathfrak{M}_o}$, turns out to be completely vague outside the observable domain $|\mathfrak{M}_o|$. The exact content of this claim is expressed by the following statements:

(1) $$\bigcap_{\mathfrak{M} \in N_{\mathfrak{M}_o}} t^{\mathfrak{M}} \subset |\mathfrak{M}_o|;$$

(2) $$\bigcap_{\mathfrak{M} \in N_{\mathfrak{M}_o}} (|\mathfrak{M}| - t^{\mathfrak{M}}) \subset |\mathfrak{M}_o|.$$

The only objects that belong to every interpretation of t, or that belong to none, are some observable things. All theoretical entities belong to the area of vagueness of this predicate. Let us now assume that t is to refer to theoretical entities only. This amounts to imposing on its interpretation the additional requirement:

$$t^{\mathfrak{M}} \cap |\mathfrak{M}_o| = \varnothing, \qquad \text{for any} \qquad \mathfrak{M} \in N_{\mathfrak{M}_o}.$$

In this case t's vagueness becomes even more striking. The statements (1) and (2) now reduce to

(3) $$\bigcap_{\mathfrak{M} \in N_{\mathfrak{M}_o}} t^{\mathfrak{M}} = \varnothing;$$

(4) $$\bigcap_{\mathfrak{M} \in N_{\mathfrak{M}_o}} (|\mathfrak{M}| - t^{\mathfrak{M}}) = |\mathfrak{M}_o|.$$

The case of a definitional extension deserves special attention. T is here obtained from T_o by adding to it a definition of t (in terms of $o_1, ..., o_n$), say:

$$(D_t) \qquad \forall x(t(x) \leftrightarrow \alpha_o(x)),$$

where $\alpha_o(x)$ belongs to L_o. The general consequences pointed out above apply, of course, to this case as well. But here some further observations may be noted. To state them, we shall distinguish two types of theoretical terms: those which refer to some observable things, and those which refer to unobservables only. Let us take a theoretical term t of the first type and assume that some element of $|\mathfrak{M}_o|$ satisfies in $\mathfrak{M}_o$ the definiens of D_t. Then, for an arbitrary set U including $|\mathfrak{M}_o|$, there is a structure $\mathfrak{M}$ in $N_{\mathfrak{M}_o}$ such that $|\mathfrak{M}| = U$ and $U - |\mathfrak{M}_o| \subset t^{\mathfrak{M}}$. Analogously, if some element of $|\mathfrak{M}_o|$ does not satisfy in $\mathfrak{M}_o$ the definiens of D_t, then for an arbitrary set U including $|\mathfrak{M}_o|$, there is a structure $\mathfrak{M}$ in $N_{\mathfrak{M}_o}$ such that $|\mathfrak{M}| = U$ and $t^{\mathfrak{M}} \cap (U - |\mathfrak{M}_o|) = \varnothing$. Thus, among the interpretations of a theoretical term t of the first type there always will be interpretations which include all the unobservable elements of a given domain, and interpretations which include none of them – whatever these elements might be. If t is a theoretical term of the second type, we may assume that no elements of $|\mathfrak{M}_o|$ satisfy in $\mathfrak{M}_o$ the definiens of D_t. Then, for an arbitrary set U including $|\mathfrak{M}_o|$, there always will be a structure $\mathfrak{M}$ in $N_{\mathfrak{M}_o}$ such that $|\mathfrak{M}| = U$ and $t^{\mathfrak{M}} = \varnothing$. Thus, among the interpretations of a theoretical term t of the second type there always will be the empty one.[3]

All these observations, however fragmentary and incomplete, help us to realize how comprehensive the class $N_{\mathfrak{M}_o}$ is bound to be and how indeterminate, in effect, the interpretation of the term t will remain. Can a theoretical term interpreted in such a way fulfil its scientific functions? One line of argumentation on this issue might run as follows. Though highly indeterminate, the interpretation of a theoretical term t is by no means arbitrary. The class of its interpretations is restricted to sets which have certain structural properties and which bear certain structural relations to interpretations of other terms. Owing to these restrictions, there may appear sentences which (1) involve term t in an essential way and, at the same time, (2) belong to

empirically decidable statements. These are notions that may be expli-
cated in different ways. The following is one of such explications. A
sentence α of language L contains term t essentially if, roughly
speaking, its truth value depends on the interpretation of t. More
precisely:

α contains t essentially iff for some $\mathfrak{M}$ and $\mathfrak{N}$,

(i) $\mathfrak{M}|_o = \mathfrak{N}|_o \in \mathrm{Mod}\,(T_o)$;

(ii) $\mathfrak{M} \in \mathrm{Mod}\,(\alpha)$ and $\mathfrak{N} \in \mathrm{Mod}\,(\sim\alpha)$.

To explain the notion of empirical decidability, I shall assume that the
set T is non-creative with respect to T_o; it may then be identified with
the set of meaning postulates in L. On this assumption, the concept of
empirical decidability may be defined as follows. A sentence α of
language L is empirically decidable if, loosely speaking, it is synthetic
and its truth value is invariant over some class $N_{\mathfrak{M}_o}$. In a formal
notation:

α is empirically decidable iff
(i) $\alpha \notin \mathrm{Cn}\,(T)$ and $\sim\alpha \notin \mathrm{Cn}\,(T)$;

(ii) for some $\mathfrak{M}_o \in \mathrm{Mod}\,(T_o)$,

$N_{\mathfrak{M}_o} \subset \mathrm{Mod}\,(\alpha)$ or $N_{\mathfrak{M}_o} \subset \mathrm{Mod}\,(\sim\alpha)$.

Now, which exactly sentences involving term t in an essential way will
belong to the class of empirically decidable statements depends, of
course, on what the postulates for t are like. There is only one
suggestion I want to make in this matter. It is my contention that the
postulates for a theoretical predicate t cannot consist of a definition of
t in terms of observational predicates $o_1, ..., o_n$, if the class of empirically decidable statements which contain t in an essential way is to
include those statements about t which are actually asserted, or put
forward, by the scientists. As an example of more adequate postulates
for a theoretical term t (of the second type), I would rather mention a
postulate of the following kind:

(P_t) $\forall x(\exists y(\beta_o(x, y) \wedge t(y)) \leftrightarrow \alpha_o(x))$,

where $\beta_o(x, y)$ and $\alpha_o(x)$ belong to L_o. It is a type of postulate proposed and discussed by several authors. The discussion provides some arguments in support of the above suggestion.[4]

Discussing the problem of identifiability in extended domains, we have restricted ourselves to some particular case of that general question and some particular interpretation of the case considered. The case has been that of extending an observational language by means of certain theoretical terms, and its interpretation has been based on the assumptions of Semantical Empiricism. The problem, however, is a more general one – not confined to that particular case and interpretation. The general schema outlined above may well apply to some other kinds of a language's extension, and their interpretation may be based on different semantical conceptions. With regard to such applications, our problem may need some reformulation. The definition of class $N_{\mathfrak{M}_o}$ may involve certain additional assumptions concerning the characterization of the extended universe and/or the interpretation of the old terms within that universe. Both may be thought of as determined not only by postulates, but also by some direct methods. In the extreme case, the domain of the extended structures will be fixed uniquely: all members of $N_{\mathfrak{M}_o}$ will be assumed to have a common domain:

$$|\mathfrak{M}| = \mathbf{U}, \qquad \text{for any} \qquad \mathfrak{M} \in N_{\mathfrak{M}_o}.$$

One of the plausible ways in which the interpretation of the old terms might be determined in the extended domains has been suggested above, when discussing the interpretation of the observational predicates. Such additional restrictions are bound to affect our conclusions concerning the degree of identifiability (or rather non-identifiability) of the terms considered. An examination of these possibilities seems worth undertaking.

University of Warsaw

NOTES

[1] The identity predicate might be treated along the same lines. Within the observable domain $|\mathfrak{M}_o|$ ' = ' might be interpreted as identity, outside $|\mathfrak{M}_o|$ – as any relation satisfying the axioms of identity.

[2] See e.g. Winnie (1967).
[3] These facts can be easily shown provided the structures for L are not restricted to normal ones, i.e. structures with standard interpretation of identity. Identity in such structures might be handled in the way mentioned before.
[4] See e.g. Przełęcki (1976).

BIBLIOGRAPHY

Przełęcki, M.: 1976, 'Interpretation of Theoretical Terms: In Defence of an Empiricist Dogma' in M. Przełęcki *et al.* (eds.), *Formal Methods in the Methodology of Empirical Sciences*, D. Reidel, Dordrecht – Ossolineum, Wrocław.
Winnie, J. A.: 1967, 'The Implicit Definition of Theoretical Terms', *The British Journal for the Philosophy of Science* **18**, 223–229.

VEIKKO RANTALA

PREDICTION AND IDENTIFIABILITY

There are many possible variants of the notion of determinism which can be connected with the logical properties of empirical theories. There has of course been a great deal of discussion about determinism in the philosophical literature but only a few attempts at a precise formulation of this notion. Noteworthy formulations can nevertheless be found in Montague (1961) and in Wójcicki (1973).

It is obvious that the notion of predictability is closely connected with the notion of determinism. As to identifiability, it seems that this concept as it is used in econometrics and social sciences does not mean, from the logical point of view, any specific kind of definability.

I shall first formulate a number of definitions which lead to a formulation of the notion of determinism which slightly differs from those earlier formulations. My starting point is closely related to the question of the value of identifiability in connection with predictions concerning the behavior of the observables described by a theory. This question is important, e.g. in econometrics (see Hurwicz, 1962). The following question may arise concerning the mathematical 'model' (theory) of a phenomenon: Is it possible to use the 'model' to obtain predictions about the future behavior of the observables on the basis of their observed past history? Here I am presupposing some sort of dichotomy between the observational terms and the theoretical terms (in some suitable language), although the definitions to be given below are independent of the question whether this kind of dichotomy can always be adequately carried out in practice.

Thus our problem is connected with the more general problem concerning the indispensability of theoretical concepts, since we are asking whether the theoretical terms of a theory can be, in a sense, passed by when predictions are made about its observables.

Second, I shall formulate a general notion of identifiability and study the relationship between these two notions.

Butts and Hintikka (eds.), Basic Problems in Methodology and Linguistics, 91–102.

It seems that there are hardly any important theories to which any 'absolute' notion of determinism could be applied. Hence I shall concentrate in the following on certain 'relativized' notions.

Let us consider a mathematical 'model' of an empirical theory. I am considering here such mathematical 'models' which are the most usual in non-stochastic empirical sciences. It seems, however, that it is not too difficult to generalize the following description of a theory as well as the following definitions so that they would be applicable also to stochastic theories. But this is not attempted here. Suppose that the 'model' includes a set of equations, or several sets of equations, which describe a behavior pattern. These equations contain certain individual variables which may be relativized to certain kinds of objects. These are the empirical objects whose behavior is described by the 'model', or other objects which form the relevant domains of empirical functions involved in the 'model'. The sets of equations may therefore be conditional in the sense which appears from the explanations given below.

Suppose that this kind of theory can be formalized in a language $L = L(\gamma, \zeta, \eta)$ where γ is a set of non-logical constants including, e.g. the set of constants needed for the mathematical basis of the theory and a set π of monadic predicate symbols, $\zeta = \{f_1, ..., f_n\}$ is a specified set of function symbols such that f_i is k_i-ary $(i = 1, ..., n)$, and $\eta = \{g_1, ..., g_m\}$ is another set of function symbols such that g_i is h_i-ary $(i = 1, ..., m)$. Individual constants are considered as 0-ary function symbols. I shall refer to the members of $\gamma \cup \zeta$ as observational terms and to the members of η as theoretical terms. L can be a first-order elementary language but this is not an essential requirement. We may assume only the usual concept of model in the model-theoretical sense of the word. Let T be the formalized theory in question.

Models (structures) for L are denoted by $\mathfrak{M} = \langle |\mathfrak{M}|, \gamma^{\mathfrak{M}}, \zeta^{\mathfrak{M}}, \eta^{\mathfrak{M}} \rangle$, etc., where $|\mathfrak{M}|$ is the domain of $\mathfrak{M}$ and $\gamma^{\mathfrak{M}}, \zeta^{\mathfrak{M}}, \eta^{\mathfrak{M}}$ are the sets of the interpretations of the constants of γ, ζ, η, respectively. If c is a constant, its interpretation is likewise denoted by $c^{\mathfrak{M}}$. For convenience, we shall write ' $\varphi^{\mathfrak{M}} = \varphi^{\mathfrak{N}}$' if $c^{\mathfrak{M}} = c^{\mathfrak{N}}$ for all $c \in \varphi$.

Since I want to be able to relativize the individual variables occurring in the behavior pattern, I suppose that π consists of those monadic predicate symbols which express these restrictions. This assumption motivates the following definitions. I do not assume any exact knowledge of the restrictions in general, that is, of the possible non-monadic or non-atomic restrictions. Since in many cases it might be adequate to take also such restrictions into account, the following definitions may somewhat simplify the matter. However, there are many theories where either all the restrictions are monadic or the non-monadic or non-atomic restrictions are such that it obviously does not diminish the adequacy of the definitions if they are neglected. On the other hand, the definitions can be modified so as to take such restrictions into account.

As an example, take the following behavior pattern which is included in the axiomatization of Newtonian particle mechanics in Simon (1947) and (1970) (cf. also Rantala, 1974):

If p is in P and t is in T,   then

$$\sum_{p \in P} m(p) D^2 s(p, t) = 0,$$

$$\sum_{p \in P} m(p) D^2 s(p, t) \times s(p, t) = 0.$$

Here P and T denote sets, m denotes a real-valued function (mass function), and s is a vector function (position function) with the second derivative $D^2 s$ (with respect to t). If this theory is formalized, we can take $\pi = \{P, T\}$, $\zeta = \{s\}$, and $\eta = \{m\}$. According to the axiomatization, if $\mathfrak{M}$ is a given model of the theory, we are interested in the values of $s^{\mathfrak{M}}$ only at the points $\langle p, t \rangle \in P^{\mathfrak{M}} \times T^{\mathfrak{M}}$ and in the values of $m^{\mathfrak{M}}$ only at the points $p \in P^{\mathfrak{M}}$.

Let $\pi = \delta \cup \varepsilon$, $\delta = \{P_{11}, ..., P_{1k_1}; ...; P_{n1}, ..., P_{nk_n}\}$ where the P_{ij} are such monadic predicate symbols (not necessarily distinct) of π that P_{ij} expresses the mentioned kind of restriction for the jth variable in f_i. Let $\mathfrak{M}$ be a model for L and $O_i^{\mathfrak{M}} = P_{i1}^{\mathfrak{M}} \times \cdots \times P_{ik_i}^{\mathfrak{M}}$ $(i = 1, ..., n)$. $H(O_1^{\mathfrak{M}}, ..., O_n^{\mathfrak{M}}) = \langle f_1^{\mathfrak{M}} \mid O_1^{\mathfrak{M}}, ..., f_n^{\mathfrak{M}} \mid O_n^{\mathfrak{M}} \rangle$ (where $f_i^{\mathfrak{M}} \mid O_i^{\mathfrak{M}}$ is the restriction of $f_i^{\mathfrak{M}}$ to $O_i^{\mathfrak{M}}$) is called the *observational history* of $\mathfrak{M}$. It gives the partial interpretations of the f_i $(i = 1, ..., n)$ restricted to the $O_i^{\mathfrak{M}}$. If $H(O_1^{\mathfrak{M}}, ..., O_n^{\mathfrak{M}}) = H(O_1^{\mathfrak{N}}, ..., O_n^{\mathfrak{N}})$ where $\mathfrak{M}$ and $\mathfrak{N}$ are models for L, then $\mathfrak{M}$ and $\mathfrak{N}$ are (observationally) *equivalent*, $\mathfrak{M} \sim \mathfrak{N}$. Let $S_1^{\mathfrak{M}}, ..., S_n^{\mathfrak{M}}$

be subsets of $O_1^{\mathfrak{M}}, ..., O_n^{\mathfrak{M}}$, respectively. $H(S_1^{\mathfrak{M}}, ..., S_n^{\mathfrak{M}}) = \langle f_1^{\mathfrak{M}} \mid S_1^{\mathfrak{M}}, ..., f_n^{\mathfrak{M}} \mid S_n^{\mathfrak{M}} \rangle$ is an *observational subhistory* of $\mathfrak{M}$. If $\mathfrak{M} \sim \mathfrak{N}$, then $O_i^{\mathfrak{M}} = O_i^{\mathfrak{N}}$ $(i = 1, ..., n)$, hence $P_{ij}^{\mathfrak{M}} = P_{ij}^{\mathfrak{N}}$ (for all $p_{ij} \in \delta$). If $H(S_1^{\mathfrak{M}}, ..., S_n^{\mathfrak{M}}) = H(S_1^{\mathfrak{N}}, ..., S_n^{\mathfrak{N}})$, then $S_i^{\mathfrak{M}} = S_i^{\mathfrak{N}}$ $(i = 1, ..., n)$. We say that $\mathfrak{N}$ is an *observational enlargement* of $\mathfrak{M}$, $\mathfrak{N} > \mathfrak{M}$, iff $H(O_1^{\mathfrak{M}}, ..., O_n^{\mathfrak{M}})$ is an observational subhistory of $H(O_1^{\mathfrak{N}}, ..., O_n^{\mathfrak{N}})$.

Let $\mathfrak{M}$ be a model of a theory T and let $S_i^{\mathfrak{M}} \subset O_i^{\mathfrak{M}}$ $(i = 1, ..., n)$ be finite. Define

$$\mathscr{F}(S_1^{\mathfrak{M}}, ..., S_n^{\mathfrak{M}}) = \{\mathfrak{N} \in \mathrm{Mod}\,(T) \colon H(S_1^{\mathfrak{N}}, ..., S_n^{\mathfrak{N}})$$

$$= H(S_1^{\mathfrak{M}}, ..., S_n^{\mathfrak{M}})$$

$$\text{for some } S_i^{\mathfrak{N}} \subset O_i^{\mathfrak{N}} \qquad (i = 1, ..., n)\}.$$

Let $\mathfrak{R}$ be a mapping from $\mathrm{Mod}\,(T)$ which assigns to each $\mathfrak{M} \in \mathrm{Mod}\,(T)$ a subset of $\mathrm{Mod}\,(T)$.

DEFINITION 1. $\mathscr{F}(S_1^{\mathfrak{M}}, ..., S_n^{\mathfrak{M}})$ is *observationally deterministic* (in short, *o.d.*) *relative to* $\mathfrak{R}$ iff for every $\mathfrak{N} \in \mathscr{F}(S_1^{\mathfrak{M}}, ..., S_n^{\mathfrak{M}}) \cap \mathfrak{R}(\mathfrak{M})$, $\mathfrak{N} \sim \mathfrak{M}$.

DEFINITION 2. T is o.d. relative to $\mathfrak{R}$ (*actually*) iff every $\mathfrak{M} \in \mathrm{Mod}\,(T)$ has a finite $S_i^{\mathfrak{M}} \subset O_i^{\mathfrak{M}}$ $(i = 1, ..., n)$ such that $\mathscr{F}(S_1^{\mathfrak{M}}, ..., S_n^{\mathfrak{M}})$ is o.d. relative to $\mathfrak{R}$.

DEFINITION 3. T is o.d. relative to $\mathfrak{R}$ (*potentially*) iff for every $\mathfrak{N} \in \mathrm{Mod}\,(T)$ there is a $\mathfrak{M} > \mathfrak{N}$, $\mathfrak{M} \in \mathrm{Mod}\,(T)$, with a finite $S_i^{\mathfrak{M}} \subset O_i^{\mathfrak{M}}$ $(i = 1, ..., n)$ such that $\mathscr{F}(S_1^{\mathfrak{M}}, ..., S_n^{\mathfrak{M}})$ is o.d. relative to $\mathfrak{R}$.

Intuitively (but somewhat loosely) speaking, the set $O_i^{\mathfrak{M}}$ consists simply of those sequences of individuals in which we are interested in connection with the observational function $f_i^{\mathfrak{M}}$. The observational history tells us what can be observed in the corresponding model. This history concerns those aspects of our individuals which are expressed by the f_i. Two models are observationally equivalent when they agree within this observational history of theirs. The sets $S_i^{\mathfrak{M}}$ may be thought of as restricting us to actual observations in a model. $\mathscr{F}(S_1^{\mathfrak{M}}, ..., S_n^{\mathfrak{M}})$ picks out those models of the theory which agree with $\mathfrak{M}$ concerning these actual observations. A theory could be thought of as being observationally deterministic in a strict, non-relativized sense if the agreement with respect to these actual observations between any two of its models $\mathfrak{M}$ and $\mathfrak{N}$ guarantees the observational equivalence of $\mathfrak{M}$ and $\mathfrak{N}$ (i.e. guarantees that they agree with respect to all possible observations), and potentially o.d. if each of its models $\mathfrak{N}$ can be 'enlarged' so as to reach this situation with respect to any other model. (Notice that 'observational enlargement' does not necessarily mean 'extension' in the usual model-theoretical sense.) Because of the way in which $\mathfrak{F}$ is defined these two strict notions would, however, be so strong that they are relativized to a set of models of the theory. This set will depend on the model under consideration. This relativization also makes these concepts flexible enough to take into account many different kinds of o.d. properties which theories may have.

In these definitions no reference is made to time. Hence they are supposed to cover postdictions as well as predictions. I also try to conform to actual prediction situations where predictions are not attempted on the basis of an infinite number of observations or perhaps on the basis of only one observation. I shall not consider the question whether these definitions correspond to what is generally meant (if any such established usage exists) by predictability in actual sciences (as far as 'observable' properties are concerned). However, later we shall see that in connection with certain particular theories they (when suitably relativized) do capture familiar ideas.

Definition 3 could be considered as an explication of the idea that we can in 'favourable circumstances' make predictions concerning the observable properties on the basis of the theory which is supposed to describe the type of phenomenon in question. When it applies, it is not the type of phenomenon itself (reflected by the form of the theory) which may prevent us from making such predictions, only the lack of sufficiently extensive and adequate observations.

Let $\varepsilon = \{R_{11}, ..., R_{1h_1}; ...; R_{m1}, ..., R_{mh_m}\}$ whose members are such predicate symbols of π (not necessarily distinct from each other or from the P_{ij}) that R_{ij} expresses the restriction for the jth variable in g_i. We do not exclude the possibility that there are no restrictions for some or all variables in a f_i, g_i. If this is the case, e.g. for the jth variable in f_i, by $P_{ij}^{\mathfrak{M}}$ we mean $|\mathfrak{M}|$, and similarly for other cases. If $\mathfrak{M}$ is a model for L, denote $Q_i^{\mathfrak{M}} = R_{i1}^{\mathfrak{M}} \times \cdots \times R_{ih_i}^{\mathfrak{M}}$ $(i = 1, ..., m)$. Let $U_i^{\mathfrak{M}}$ $(i = 1, ..., n)$ be proper subsets of the $O_i^{\mathfrak{M}}$, respectively. $H(U_i^{\mathfrak{M}}, ..., U_n^{\mathfrak{M}})$

is a *determining subhistory* of $\mathfrak{M}$ iff for every $\mathfrak{N} \in \mathrm{Mod}\,(T)$ the following holds: For every subhistory $H(U_1^{\mathfrak{N}}, ..., U_n^{\mathfrak{N}})$ of $\mathfrak{N}$, if $H(U_1^{\mathfrak{N}}, ..., U_n^{\mathfrak{N}}) = H(U_1^{\mathfrak{M}}, ..., U_n^{\mathfrak{M}})$, $g_i^{\mathfrak{N}} \mid Q_i^{\mathfrak{N}} = g_i^{\mathfrak{M}} \mid Q_i^{\mathfrak{M}}$ $(i = 1, ..., m)$, and $\gamma^{\mathfrak{N}} = \gamma^{\mathfrak{M}}$, then $\mathfrak{N} \sim \mathfrak{M}$.

In the following definition it is supposed that every model of T has a determining subhistory. For each $\mathfrak{M} \in \mathrm{Mod}\,(T)$ choose a fixed determining subhistory $H(U_1^{\mathfrak{M}}, ..., U_n^{\mathfrak{M}})$. Denote by $\mathcal{H}$ the mapping from $\mathrm{Mod}\,(T)$ for which

$$\mathcal{H}(\mathfrak{M}) = \{\mathfrak{N} \in \mathrm{Mod}\,(T) : H(U_1^{\mathfrak{N}}, ..., U_n^{\mathfrak{N}})$$
$$= H(U_1^{\mathfrak{M}}, ..., U_n^{\mathfrak{M}}) \quad \text{for some}$$
$$U_1^{\mathfrak{N}} \subset O_i^{\mathfrak{N}} \, (i = 1, ..., n), \quad \text{and} \quad \gamma^{\mathfrak{N}} = \gamma^{\mathfrak{M}}\}.$$

DEFINITION 4. If in Definitions 2 and 3, $\mathfrak{R} = \mathcal{H}$, T is *o.d. in the determining subhistories* in question (actually or potentially, respectively).

Definition 4 corresponds, with suitably chosen $U_i^{\mathfrak{M}}$'s of a given $\mathfrak{M}$, in a sense to the case where by means of a finite number of observations the set of all those possible values of the observational functions $f_i^{\mathfrak{M}}$ which satisfy the equations are determined. The correspondence obtains in the sense that if any set of values is substituted for the 'occurrences' of the observational terms in the equations of the mathematical 'model' in question, it can be decided whether these values satisfy the equations. This does not mean that we then know the whole observational history of $\mathfrak{M}$. It only means, e.g. that if we know a certain subhistory of $\mathfrak{M}$, then the rest of the history of $\mathfrak{M}$ is uniquely determined (cf. Theorem 5, below).

We say that $f_1, ..., f_k$ $(k < n)$ are the *predetermined functions* of T and $f_{k+1}, ..., f_n$ are the *endogenous functions* of T iff for every pair $\mathfrak{M}, \mathfrak{N} \in \mathrm{Mod}\,(T)$ the following holds: If $f_i^{\mathfrak{M}} \mid O_i^{\mathfrak{M}} = f_i^{\mathfrak{N}} \mid O_i^{\mathfrak{N}}$ $(i = 1, ..., k)$, $g_i^{\mathfrak{M}} \mid Q_i^{\mathfrak{M}} = g_i^{\mathfrak{N}} \mid Q_i^{\mathfrak{N}}$ $(i = 1, ..., m)$, and $\gamma^{\mathfrak{M}} = \gamma^{\mathfrak{N}}$, then $f_i^{\mathfrak{M}} \mid O_i^{\mathfrak{M}} = f_i^{\mathfrak{N}} \mid O_i^{\mathfrak{N}}$ $(i = k+1, ..., n)$.

Denote by $\mathcal{P}$ the mapping from $\mathrm{Mod}\,(T)$ for which

$$\mathcal{P}(\mathfrak{M}) = \{\mathfrak{N} \in \mathrm{Mod}\,(T) : f_i^{\mathfrak{N}} \mid O_i^{\mathfrak{N}} = f_i^{\mathfrak{M}} \mid O_i^{\mathfrak{M}} \quad (i = 1, ..., k)$$
$$\text{and} \quad \gamma^{\mathfrak{N}} = \gamma^{\mathfrak{M}}\},$$

where $f_1, ..., f_k$ are the predetermined functions of T. So it is assumed here that T has predetermined and endogenous functions which have been specified in advance.

DEFINITION 5.   If in Definitions 2 and 3, $\mathfrak{R} = \mathfrak{P}$, T is *o.d. in the predetermined functions* $f_1, ..., f_k$ (actually or potentially, respectively).

The above definition of predetermined and endogenous functions corresponds to the distinction made by econometricians and social scientists between independent and dependent variables, or predetermined and endogenous variables. The definition is also closely connected with the identification of the endogenous variables by means of the predetermined variables and the theoretical terms. It says that if the partial interpretations of the predetermined functions (that is, their interpretations in the respective sets $O_i^{\mathfrak{M}}$) are known, as well as the partial interpretations (in the $Q_i^{\mathfrak{M}}$) of the theoretical terms, then the partial interpretations of the endogenous functions are uniquely determined. (The definition is also related to the distinction made in system theory between input and output variables.) However, as Definition 5 shows, we are here interested in the question whether such identification of the endogenous functions is possible without the exact knowledge of the theoretical terms (cf. also the notion of conditional determinism in Montague (1961)).

There obtains an interesting connection between the notion of observational determinism and the notion of conditional determinism in the sense of Montague (1961) (when L and T are chosen suitably). I restrict myself to the case where π consists of just one monadic predicate symbol R. But the following theorem holds also for the more general theories considered by Montague. Notice that in a standard model $\mathfrak{M}$ of T in the sense of Montague (1961), $R^{\mathfrak{M}}$ is the set of real numbers and the mathematical terms are interpreted in the usual way. According to Montague, the *state* of the *history* $\langle \zeta^{\mathfrak{M}}, \eta^{\mathfrak{M}} \rangle$ at t, where $\mathfrak{M}$ is standard and $t \in R^{\mathfrak{M}}$, is

$$st_{\mathfrak{M}}(t) = \langle f_1^{\mathfrak{M}}(t), ..., f_n^{\mathfrak{M}}(t), g_1^{\mathfrak{M}}(t), ..., g_m^{\mathfrak{M}}(t) \rangle.$$

T is *deterministic in* η in the sense of Montague iff for all standard models $\mathfrak{M}, \mathfrak{N} \in \text{Mod}(T)$ and for all $t_0, t \in R^{\mathfrak{M}}$, if $\eta^{\mathfrak{M}} = \eta^{\mathfrak{N}}$ and $st_{\mathfrak{M}}(t_0) = st_{\mathfrak{N}}(t_0)$, then $st_{\mathfrak{M}}(t) = st_{\mathfrak{N}}(t)$.

Define now a mapping $\mathscr{S}$ from Mod (T) as follows:

$$\mathscr{S}(\mathfrak{M}) = \{\mathfrak{N} \in \text{Mod}\,(T): \mathfrak{N} \text{ is standard and } \eta^{\mathfrak{N}} = \eta^{\mathfrak{M}}\}$$

if $\mathfrak{M}$ is standard,

$$\mathscr{S}(\mathfrak{M}) = \varnothing \text{ if } \mathfrak{M} \text{ is not standard.}$$

THEOREM 1. *If T is deterministic in η, then T is o.d. relative to $\mathscr{S}$ (actually).*

Proof. Let $\mathfrak{M} \in \text{Mod}\,(T)$. If $\mathfrak{M}$ is not standard, then the condition of Definition 2 holds trivially for any $S_i^{\mathfrak{M}}$ $(i = 1, ..., n)$. Let $\mathfrak{M}$ be standard and $S_i^{\mathfrak{M}} = \{t_0\}$ $(i = 1, ..., n)$, $t_0 \in R^{\mathfrak{M}}$ arbitrary. If $\mathfrak{N} \in \mathscr{F}(S_i^{\mathfrak{M}}, ..., S_n^{\mathfrak{M}}) \cap \mathscr{S}(\mathfrak{M})$, then $f_i^{\mathfrak{N}} | \{t_0\} = f_i^{\mathfrak{M}} | \{t_0\}$ $(i = 1, ..., n)$, $\mathfrak{N}$ is standard, and $\eta^{\mathfrak{N}} = \eta^{\mathfrak{M}}$. Thus $st_{\mathfrak{N}}(t_0) = st_{\mathfrak{M}}(t_0)$ and $\eta^{\mathfrak{N}} = \eta^{\mathfrak{M}}$, whence $st_{\mathfrak{N}}(t) = st_{\mathfrak{M}}(t)$ for every $t \in R^{\mathfrak{M}}$ since T is deterministic in η. Hence for every $t \in R^{\mathfrak{M}}$, $f_i^{\mathfrak{N}} | \{t\} = f_i^{\mathfrak{M}} | \{t\}$ $(i = 1, ..., n)$, that is, $f_i^{\mathfrak{N}}(t) = f_i^{\mathfrak{M}}(t)$ $(i = 1, ..., n)$. Thus $\mathfrak{N} \sim \mathfrak{M}$. ∎

I shall now define a notion of identifiability which is closely related to (but stated in the more general framework than) the corresponding notion used by econometricians and social scientists as far as non-stochastic mathematical 'models' are concerned. If we consider stochastic 'models', this notion is obviously related to the notion of estimation.

It seems that in actual scientific practice it is often asked whether it is *possible* to find a *finite* number of observations by means of which the (restricted) values of a theoretical term could be determined uniquely (or estimated within reasonable limits of approximation).

Again, I first define a relativized concept, for this makes it possible to take into account different kinds of identifiability properties theories may have. So, let $\mathfrak{R}$ be a mapping from Mod (T) which assigns to each $\mathfrak{M} \in \text{Mod}\,(T)$ a subset of Mod (T).

DEFINITION 6. (a) g_i is *identifiable in T relative to $\mathfrak{R}$ (potentially)* iff for every $\mathfrak{M}^* \in \text{Mod}\,(T)$ there is a $\mathfrak{M} > \mathfrak{M}^*$, $\mathfrak{M} \in \text{Mod}\,(T)$, with a finite $S_i^{\mathfrak{M}} \subset O_i^{\mathfrak{M}}$ $(i = 1, ..., n)$ such that whenever $\mathfrak{N} \in \mathscr{F}(S_1^{\mathfrak{M}}, ..., S_n^{\mathfrak{M}}) \cap \mathfrak{R}(\mathfrak{M})$, then $g_i^{\mathfrak{N}} | Q_i^{\mathfrak{N}} = g_i^{\mathfrak{M}} | Q_i^{\mathfrak{M}}$.

(b) η is identifiable in T relative to $\mathcal{R}$ (potentially) iff for every $\mathfrak{M}^* \in \text{Mod}(T)$ there is a $\mathfrak{M} > \mathfrak{M}^*$, $\mathfrak{M} \in \text{Mod}(T)$, with a finite $S_i^{\mathfrak{M}} \subset O_i^{\mathfrak{M}}$ $(i = 1, ..., n)$ such that whenever $\mathfrak{N} \in \mathcal{F}(S_1^{\mathfrak{M}}, ..., S_n^{\mathfrak{M}}) \cap \mathcal{R}(\mathfrak{M})$, then $g_i^{\mathfrak{N}} \mid Q_i^{\mathfrak{N}} = g_i^{\mathfrak{M}} \mid Q_i^{\mathfrak{M}}$ for all $i = 1, ..., m$.

(a^+)–(b^+) Let $\varphi = (\gamma - \delta) \cup (\delta \cap \varepsilon)$. If in (a) or (b), $\mathcal{R} = \mathcal{I}$ such that $\mathcal{I}(\mathfrak{M}) - \{\mathfrak{N} \subset \text{Mod}(T): \varphi^{\mathfrak{N}} = \varphi^{\mathfrak{M}}\}$, g_i or η, respectively, is *identifiable in T (potentially)*.

(a^{++})–(b^{++}) If in (a)–(b^+), $\mathfrak{M}^*$ itself serves as $\mathfrak{M}$, for every $\mathfrak{M}^* \in \text{Mod}(T)$, we obtain the respective notions of *actual* identifiability.

Suppose that every $g_i \in \eta$ is identifiable in T (potentially). This implies that η is identifiable in T (potentially) if the following holds: By applying Definition $6(a^+)$ consecutively to $g_1, ..., g_m$, we can find for an arbitrary $\mathfrak{M}^* \in \text{Mod}(T)$, models $\mathfrak{M}_1, ..., \mathfrak{M}_m = \mathfrak{M}$, $\mathfrak{M}^* < \mathfrak{M}_1 < \cdots < \mathfrak{M}_m$, with the corresponding finite sets $S_i^{\mathfrak{M}_1}, ..., S_i^{\mathfrak{M}_m}$ $(i = 1, ..., n)$, such that $S_i^{\mathfrak{M}_j} \subset S_i^{\mathfrak{M}_k}$, $\varphi^{\mathfrak{M}_j} = \varphi^{\mathfrak{M}_k}$, and $g_j^{\mathfrak{M}_j} \mid Q_j^{\mathfrak{M}_j} = g_j^{\mathfrak{M}_k} \mid Q_j^{\mathfrak{M}_k}$ for $1 \leqslant j \leqslant k \leqslant m$.

This condition holds for such simple econometrical theories as were considered in Rantala (1974). Also the notion of identifiability by Definition $6(a^+)$ coincides with its namesake in that paper for this kind of theories. However, in order that the general description (that was given in the beginning of this paper) concerning the kind of theories considered here would be formally applicable to such econometrical theories, the following syntactical modification has to be done: It is supposed now that the (conditional) behavior patterns of these econometrical theories are of the form

$$(\forall t)(P(t) \rightarrow F(a_1, ..., a_m; x_1(t), ..., x_n(t))),$$

where $\pi = \{P\}$, $\zeta = \{x_1, ..., x_n\}$, and $\eta = \{a_1, ..., a_m\}$. F is a set of equations, $x_1, ..., x_n$ are unary function symbols, and $a_1, ..., a_m$ are individual constants (coefficients). P is a monadic predicate symbol which intuitively can be thought to correspond to time or, more loosely, to a set indexing the 'observations' $\langle x_1(t), ..., x_n(t) \rangle$. All the definitions and results of the mentioned paper still hold with obvious modifications.

A suitably relativized notion of identifiability coincides with implicit definability. Suppose, for simplicity, that $\eta = \{g\}$. Let L_0 be the reduction of L to the vocabulary of $\gamma \cup \zeta$. If $\mathfrak{M}$ is a model for L, its reduct to

L_0 is denoted by $\mathfrak{M} \mid L_0$. Let $\mathfrak{R} = \mathfrak{E}$ where

$$\mathfrak{E}(\mathfrak{M}) = \{\mathfrak{N} \in \text{Mod}\,(T): \mathfrak{N} \mid L_0 = \mathfrak{M} \mid L_0\}.$$

g is implicitly definable in T iff for every $\mathfrak{M} \in \text{Mod}\,(T)$, whenever $\mathfrak{N} \in \mathfrak{E}(\mathfrak{M})$, then $g^{\mathfrak{N}} = g^{\mathfrak{M}}$ (hence $\mathfrak{N} = \mathfrak{M}$). If Beth's Theorem holds for L (as e.g. if L is a first order elementary language), this condition is equivalent with the explicit definability of g and the following theorem yields the connection between explicit definability and identifiability. There obtains also a similar connection between identifiability and conditional definability.

THEOREM 2. *Suppose T does not pose any restrictions for the variables in g. Then g is implicitly definable in T iff g is identifiable in T relative to $\mathfrak{E}$ (actually).*

Proof. Follows immediately from the fact that, given any $\mathfrak{M} \in \text{Mod}\,(T)$ and any $S_i^{\mathfrak{M}} \subset O_i^{\mathfrak{M}}$ ($i = 1, ..., n$), $\mathfrak{E}(\mathfrak{M}) \subset \mathscr{F}(S_1^{\mathfrak{M}}, ..., S_n^{\mathfrak{M}})$. ■

Identifiability is a stronger notion than observational determinism in the following sense.

THEOREM 3. *Let $f_1, ..., f_k$ be the predetermined functions of T. If η is identifiable in T (potentially or actually), then T is o.d. in the predetermined functions $f_1, ..., f_k$ (potentially or actually, respectively).*

Proof. Consider only the potential case. Let $\mathfrak{R} = \mathscr{P}$ where $f_1, ..., f_k$ are the predetermined functions of T, and let $\mathfrak{N}$ be an arbitrary model of T. If η is identifiable in T (potentially), we can find a $\mathfrak{M} > \mathfrak{N}$, $\mathfrak{M} \in \text{Mod}\,(T)$, with a finite $S_i^{\mathfrak{M}} \subset O_i^{\mathfrak{M}}$ ($i = 1, ..., n$), such that the condition of Definition 6(b$^+$) is satisfied. Let $\mathfrak{M}' \in \mathscr{F}(S_1^{\mathfrak{M}}, ..., S_n^{\mathfrak{M}}) \cap \mathscr{P}(\mathfrak{M})$. Since $\mathscr{P}(\mathfrak{M}) \subset \mathscr{I}(\mathfrak{M})$, $\mathfrak{M}' \in \mathscr{F}(S_1^{\mathfrak{M}}, ..., S_n^{\mathfrak{M}}) \cap \mathscr{I}(\mathfrak{M})$. Thus $g_i^{\mathfrak{M}'} \mid Q_i^{\mathfrak{M}'} = g_i^{\mathfrak{M}} \mid Q_i^{\mathfrak{M}}$ ($i = 1, ..., m$) by Definition 6(b$^+$). Since $\mathfrak{M}' \in \mathscr{P}(\mathfrak{M})$, $f_i^{\mathfrak{M}'} \mid O_i^{\mathfrak{M}'} = f_i^{\mathfrak{M}} \mid O_i^{\mathfrak{M}}$ ($i = 1, ..., k$) and $\gamma^{\mathfrak{M}'} = \gamma^{\mathfrak{M}}$. By the definition of predetermined and endogenous functions, $f_i^{\mathfrak{M}'} \mid O_i^{\mathfrak{M}'} = f_i^{\mathfrak{M}} \mid O_i^{\mathfrak{M}}$ ($i = k+1, ..., n$). Hence $\mathfrak{M}' \sim \mathfrak{M}$. ■

Thus the identifiability (in the sense of Definition 6(b$^+$) or 6(b^{++}) of the theoretical terms makes it possible to obtain predictions concerning the observable properties on the basis of a finite number of observations. It can be shown that explicit or conditional definability is not sufficient in general.

Finally, let us consider the two familiar kinds of theories mentioned in the above. The following theorem is obvious.

THEOREM 4. *Let T be an econometrical theory of the kind specified above.*

   (i) *T is o.d. in the predetermined functions (potentially).*

   (ii) *If A_{n_T} is the sentence which says that in every model of T there is the maximal number of mutually independent observations (cf. Rantala, 1974), then $T \cup \{A_{n_T}\}$ is o.d. in the predetermined functions (actually).*

This theorem, which is due to the fact that the parameters (theoretical functions) are individual constants, can be considered as an explication of the known fact that on the basis of this kind of theory it is possible to make predictions about observables, even though the parameters are not identifiable. But, as is remarked in Hurwicz (1962), it is not always possible to make such predictions if some of the parameters are not constant unless these parameters are identifiable. It is supposed in Hurwicz (1962) that the parameters will change in a known way, that is, if the values of the parameters in some time interval before the change are known, then their values after the change can be calculated. This assumption may guarantee the possibility of identification of the parameters. If it does, then by Theorem 3 it also guarantees the possibility of prediction. In view of Theorem 3, however, it is clear that in some cases this assumption can guarantee the latter possibility but not the former.

Also the following result is obvious. Consider a theory of Newtonian particle mechanics referred to in the above example. A model of this theory which includes n 'particles' $p_1, ..., p_n$ has a determining subhistory in the following sense: If the 'positions' of $p_1, ..., p_{n-1}$ are known at any 'time instant', as well as the 'masses' of $p_1, ..., p_n$, the position of p_n is uniquely determined at any time instant.

THEOREM 5. (i) *A formalized theory of Newtonian particle mechanics (as axiomatized in Simon (1947) and (1970)) is o.d. in the determining subhistories (potentially).*

(ii) *A formalized theory of holomorphic Newtonian particle mechanics* (*see Simon*, 1970) *is o.d. in the determining subhistories* (*actually*).

This theorem follows from the fact that mass is considered constant in time. In general, it seems that observational determinism in any non-trivial relativized sense, corresponding to actual prediction situations, is such a strong property, although it is in a sense weaker than determinism defined by Montague, that only rather specific or simple kinds of theories possess it. They are either theories in which at least some of the theoretical terms are identifiable, or theories whose theoretical terms have certain other 'constancy' properties.

Academy of Finland, Helsinki, Finland

BIBLIOGRAPHY

Hurwicz, L.: 1962, 'On the Structural Form of Interdependent Systems', in E. Nagel, P. Suppes, and A. Tarski (eds.), *Logic, Methodology and Philosophy of Science*, Stanford University Press, Stanford, pp. 232–239.
Montague, R.: 1961, 'Deterministic Theories', in Washburne (ed.), *Decisions, Values, and Groups*, Vol. 2, Pergamon Press, Oxford, pp. 325–367.
Rantala, V.: 1974, 'Definability Problems in the Methodology of Science', in the *Proceedings of the Conference for Formal Methods in the Methodology of Empirical Sciences*, Warsaw 1974, D. Reidel, Dordrecht, (forthcoming).
Simon, H.: 1947, 'The Axioms of Newtonian mechanics', *Phil. Mag.*, ser. 7, 33.
Simon, H.: 1970, 'The Axiomatization of Physical Theories', *Philosophy of Science* **37**, 16–26.
Wójcicki, R.: 1973, 'Basic Concepts of Formal Methodology of Empirical Sciences', *Ajatus* **35**, 168–196.

III

FOUNDATIONS OF
PROBABILITY AND INDUCTION

TERRENCE L. FINE

AN ARGUMENT FOR COMPARATIVE PROBABILITY

1. BACKGROUND TO AND OUTLINE OF THE ARGUMENT

We undertake to argue in favor of generalizing the formal concept of probability by replacing the usual quantitative formulation by the hitherto largely ignored comparative formulation. The theory of comparative probability (CP) is a theory of statements of the forms 'event A is more probable than event B', 'events A, B are equally probable', 'event A is at least as probable as event B', having the respective symbolic representations, '$A > B$', '$A \sim B$', '$A \geqslant B$'. Comparative probability seems to have first been studied by Bernstein (1917), Keynes (1921), de Finetti (1931), and Koopman (1940). The modern era of CP studies perhaps dates from the work of Carnap (1950), Savage (1954), and Kraft *et al.* (1959). Relatively recent contributors include Scott (1964), Luce (1967, 1968), Fishburn (1970, 1975), Domotor (1969), Fine (1971a, b, 1973), Fine and Gill (1976), Fine and Kaplan (1976), Kaplan (1971, 1974), and Narins (1974).

Unfortunately in our view the goal of most of this work, with the clear exceptions of Keynes (1921), Koopman (1940), Fine (1971b, 1973), Fine and Gill (1976), Fine and Kaplan (1976), and Kaplan (1971, 1974), was to delineate a derivation and justification for the usual quantitative concept of probability. In Fine (1971b, 1973), Fine and Gill (1976), Fine and Kaplan (1976), and Kaplan (1971, 1974) an attempt was made to study CP as of interest and importance in its own right. This work revealed properties of CP that are intuitively attractive and that distinguish it from quantitative probability. These properties, discussed below, pertain to assumptions of the universal comparability of the degrees of uncertainty or chance, the possibilities of combining individual models of uncertainty and chance phenomena into a single joint model, and limits to the precision with which we can measure uncertainty or chance. Recently we have also become aware

Butts and Hintikka (eds.), Basic Problems in Methodology and Linguistics, 105–119.
Copyright © 1977 by D. Reidel Publishing Company, Dordrecht-Holland. All Rights Reserved.

of the possibilities for a frequentist interpretation of CP that extends the frequentist interpretations given to quantitative probability. This frequentist interpretation supplements the subjective and logical interpretations that have hitherto been thought of in connection with CP.

As side benefits of our efforts we uncover a new characterization of the usual quantitative probability models and may be able to reopen a question about convergence to certainty when one makes unlimited, unlinked repetitions of a random experiment that J. Bernoulli felt he had disposed of with his law of large numbers.

In large part many of the controversies affecting the foundations of probability and thence communicated to the foundations of statistics seem to us to be a result of attempts to answer too sophisticated a question too quickly as well as a legacy of operating in an arena narrowed by choices made too early in the history of probability. By turning to the logically simpler concept of comparative probability we might hope to be more successful in clarifying the nature of chance and uncertainty phenomena. Certainly if we cannot explicate comparative probability then there is little hope for an explication of the usual concept of quantitative probability.

The development of our argument requires certain definitions and theorems to which we now turn.

2.   Axioms for comparative probability

We will only consider the case of finite event collections. These event collections will be assumed to be algebras although it is not necessary to do so. The axioms we initially provide for the binary relations $>$, $\sim$, $\geqslant$ have not been well-studied. Let $\mathfrak{A}$ denote the finite event algebra, A, B, C, ..., generic elements of $\mathfrak{A}$, Ω the sample space or certain event, φ the empty set or impossible event. By $A \geqslant B$ we mean either $A > B$ or $A \sim B$.

PCP1.   (Non-triviality) $\Omega > \varphi$, $A \sim A$.

PCP2.   (Usage)   (a) $A \sim B \Rightarrow B \sim A$.
                          (b)   False $A > B$ and $B \geqslant A$.

PCP3.   (Restricted Transitivity)   False $A > B$, $B \geqslant C$, and $C \geqslant A$.

PCP4.   (Monotonicity)   (a)   False $\varphi > A$.
                           (b)   If $A \supset B > C \supset D$ then $A > D$.

PCP5.   (Pre-additivity)   If $A \cap (B \cup C) = \varphi$ then $B \geqslant C \Leftrightarrow A \cup B \geqslant A \cup C$.

PCP6.   (Dominance)   (a)   If $A \cap B = \varphi$, $A \geqslant C$ and $B \geqslant D$ then false $C \cup D > A \cup B$.
                           (b)   If also $A > C$ then false $C \cup D \geqslant A \cup B$.

LEMMA 1.   *If for all* A, $B \in \mathfrak{A}$ *either* $A \geqslant B$ *or* $B \geqslant A$ *then* PCP1–PCP6 *reduce to* $\geqslant$ *is complete, transitive and* PCP1, PCP4a, PCP5.

The usual axioms for CP attributed to B. de Finetti are the conclusion of Lemma 1. As we shall see, the assumption that $\geqslant$ is a complete order is sometimes impossible and necessitates starting with partial comparative probability (PCP).

If we wish to extend the formulation of CP or PCP to infinite algebras then additional axioms would seem desirable. Candidate axioms would include the existence of a countable order-dense set or the second countability of the order topology (Fine, 1971a; Roberts, 1973), monotone continuity (Villegas, 1964), and Archimedean axioms (Krantz *et al.*, 1971). We do not pursue this issue further.

Two elementary consequences of PCP1–PCP6 are:

(i)   False $A \supset B$, $B > A$.
(ii) $A \geqslant B \Rightarrow \bar{B} \geqslant \bar{A}$, where $\bar{A}$ denotes the complement of A.

3. CLASSIFICATION OF CP RELATIONS

A CP relation $\geqslant$ is *additive* if there exists a probability measure P on $\mathfrak{A}$ such that

$$A \geqslant B \Leftrightarrow P(A) \geqslant P(B).$$

A CP relation is *non-additive* if it is not additive.

To refine the notion of a non-additive relation we can consider:

almost additive relations $- A \geqslant B \Rightarrow P(A) \geqslant P(B)$,

weakly additive relations $- P(A) \geqslant P(B) \Rightarrow A \geqslant B$

Strictly non-additive relations $-$ not almost additive.

The existence of non-additive relations was first noticed by Kraft *et al.* [7] who asserted that:

(i)   all complete relations on four or fewer atoms were additive;
(ii) there exist almost additive complete relations between subsets of a sample space of 5 atoms;
(iii) there exist strictly non-additive complete relations between subsets of a sample space of 6 atoms.

There is no known 'fast' algorithm for generating the non-additive relations nor is it easy to determine the classification of a given relation. There appear to be large numbers of CP relations on even small algebras (Fine and Gill, 1976).

We are primarily interested in the new models for uncertainty and chance phenomena that are represented by the non-additive CP relations. A key distinction between the additive and non-additive relations is the degree to which they can be extended via the operation of forming joint relations or orders.

4.   JOINT ORDERS

If we are given individual random experiments $\{\mathscr{E}_i, i = 1, n\}$ described by their event algebras $\mathfrak{A}_i$ and CP relations $\geqslant_i$, $\mathscr{E}_i = (\mathfrak{A}_i, \geqslant_i)$, then we may ask about the existence of the joint experiment $\mathscr{E}^{(n)} = (\mathfrak{A}^{(n)}, \geqslant^{(n)})$ where $\mathfrak{A}^{(n)}, \geqslant^{(n)}$ are defined as follows. Let $A_i \in \mathfrak{A}_i$, $\overset{n}{\underset{i=1}{\mathsf{X}}} A_i$ denote the Cartesian product of coordinate events, $\alpha, \beta, \dots,$ denote subsets of $\{1, \dots, n\}$,

$$\left[\underset{i \in \alpha}{\mathsf{X}} A_i \right] = \left(\underset{i \in \alpha}{\mathsf{X}} A_i \right) \times \left(\underset{i \notin \alpha}{\mathsf{X}} \Omega_i \right),$$

and $\mathfrak{A}^{(n)}$ be the smallest algebra containing all of $\underset{i=1}{\overset{n}{X}} A_i$. The relation $\geqslant^{(n)}$ is a joint order extending $\{\geqslant_i\}$ if $\geqslant^{(n)}$ satisfies PCP1–PCP6 as a binary relation between elements of $\mathfrak{A}^{(n)}$ and for all i, for all A_i, $B_i \in \mathfrak{A}_i$,

$$[A_i] \geqslant^{(n)} [B_i] \Leftrightarrow A_i \geqslant_i B_i.$$

A joint order is exchangeable if for all i, $\mathfrak{A}_i = \mathfrak{A}$, $\geqslant_i = \geqslant$, and for all $A_i \in \mathfrak{A}$ and $\{i_j\}$ permutations of $\{j\}$, it is false that $\underset{j=1}{\overset{n}{X}} A_j >^{(n)} \underset{j=1}{\overset{n}{X}} A_{i_j}$.

A joint order is of independent type if for all disjoint index sets α, β, for all A_j, B_j, C_j, $D_j \in \mathfrak{A}_j$,

(i) False $\left[\underset{j \in \alpha}{X} A_j\right] \geqslant^{(n)} \left[\underset{j \in \alpha}{X} B_j\right]$, $\left[\underset{j \in \beta}{X} C_j\right] \geqslant^{(n)} \left[\underset{j \in \beta}{X} D_j\right]$ and

$$\left[\left(\underset{j \in \alpha}{X} B_j\right) \times \left(\underset{j \in \beta}{X} D_j\right)\right] >^{(n)} \left[\left(\underset{j \in \alpha}{X} A_j\right) \times \left(\underset{j \in \beta}{X} C_j\right)\right];$$

(ii) If also $\left[\underset{j \in \alpha}{X} A_j\right] >^{(n)} \left[\underset{j \in \alpha}{X} B_j\right]$ then false

$$\left[\left(\underset{j \in \alpha}{X} B_j\right) \times \left(\underset{j \in \beta}{X} D_j\right)\right] \geqslant^{(n)} \left[\left(\underset{j \in \alpha}{X} A_j\right) \times \left(\underset{j \in \beta}{X} C_j\right)\right].$$

A joint order is of independent and identically distributed (IID) type if it is both exchangeable and of independent type.

It is easily verified that the preceding definitions of exchangeable, independent, and IID orders agree with the usual quantitative definitions when the probability measures P_i agree with $\geqslant_i$ and in the independent case

$$P^{(n)}(A_1, ..., A_n) = \prod_{i=1}^{n} P_i(A_i) \text{ agrees with } \geqslant^{(n)}.$$

The novelty introduced by comparative probability is that there need not exist complete joint orders extending given complete individual orders.

5. CHARACTERIZATION OF THE ADDITIVE CP RELATIONS

The following theorem, proven in Fine and Kaplan (1976), characterizes the usual quantitatively induced relations as exactly the class that admits of IID extension to a complete joint order.

THEOREM 1.   *A random experiment* $\mathcal{E} = (\mathfrak{A}, \succcurlyeq)$ *with* $\mathfrak{A}$ *finite admits of extension for all finite n to an n-fold IID type complete joint experiment* $\mathcal{E}^{(n)} = (\mathfrak{A}^{(n)}, \succcurlyeq^{(n)})$ *if and only if* $\succcurlyeq$ *is an additive relation.*

The 'if' part of the theorem is established in the usual manner through $P^{(n)}$ an agreeing product measure. The 'only if' part is of interest as it asserts that for any non-additive $\succcurlyeq$ there is a largest n (possibly 1) for which there exists a complete joint description of n IID experiments each of which is described by $\succcurlyeq$.

There does not appear to us to be any argument that all situations of uncertainty or chance must admit of IID repetition. Failing such an argument the non-additive orders have a claim to usefulness as models. Note that one must distinguish here between accepting, as we do, that one can have a collection of unlinked random experiments in which the outcomes of one experiment do not influence the outcomes of another experiment and characterizing this collection of experiments as a single, completely described experiment via the model of independence.

A restatement of Theorem 1 is

THEOREM 2.   *Let* $\mathcal{U}_n$ *be the random experiment with a sample space* Ω *of n-atoms all of which are equally probable (i.e.* $A \succcurlyeq B$ *iff* $\|A\| \succcurlyeq \|B\|$*). A random experiment* $\mathcal{E} = (\mathfrak{A}, \succcurlyeq)$ *admits of a complete independent joint order with* $\mathcal{U}_n$ *for any finite n if and only if* $\succcurlyeq$ *is additive.*

The assumption that a given random experiment can be independently combined with any discrete uniformly distributed random variable leads to the device of compound lotteries successfully used by subjective probabilists.

6. COMPARABILITY OF UNCERTAINTIES

As suggested by Theorem 2, implicit in a reliance upon only the conventional additive CP models is the assumption that all uncertainties can be compared with the single scale of uncertainty generated by the discrete uniform distribution. Furthermore it is assumed that this scale can be arbitrarily refined to make comparisons illimitably precise at least in principle. The belief that all uncertainties or chance phenomena are mutually comparable is one to which we have grown accustomed when acting in a 'technical' or 'professional' capacity but one easily discomfited by naive personal introspection. While I may reasonably believe that it is more probable that a paper I am writing will be accepted by a particular journal than that it will be rejected, this in no way commits me to the in principle arbitrarily precise calibration of the strength of my belief. I feel no obligation to be able to compare the probability of acceptance with the probabilities of all subsets of outcomes of 1000 tosses of a fair coin.

The belief that not all uncertainties are comparable is common even among those who accept only the quantitative concept of probability. Examples are evident in non-Bayesian statistics, the principle of complementarity in quantum mechanics (see Fine, 1974), the difficulties in defining σ-additive measures for, say, all subsets of the reals, and realistic accounts of human decision-making.

Note, however, that we are not referring to an inability to compare uncertainties due to a lack of knowledge but rather to its opposite. It is our very knowledge about the uncertainty structure of two experiments that makes some comparisons between them impossible. To be more precise we can define quantitative upper and lower scales $\bar{s}$, $\underline{s}$ for a random CP experiment $(\mathfrak{A}, \geqslant)$ as follows:

(i) $A \geqslant B \Rightarrow \bar{s}(A) \geqslant \underline{s}(B) \geqslant 0$,

(ii) $A \cap B = \varphi \Rightarrow \bar{s}(A \cup B) \leqslant \bar{s}(A) + \bar{s}(B)$, $\underline{s}(A \cup B) \geqslant \underline{s}(A) + \underline{s}(B)$.

We note

THEOREM 3. (a) *If $\geqslant$ is complete and almost additive but not additive then there exist $\{A_i, B_i\}$ such that*

$$(\forall i)\ A_i \geqslant B_i, \quad (\exists j)\ A_j > B_j,$$

$$(\forall \omega \in \Omega)\ \|\{A_i : \omega \in A_i\}\| = \|\{B_i : \omega \in B_i\}\|.$$

(b) *If $\geqslant$ is strictly non-additive then there exists $\{A_i, B_i\}$ that*

$$(\forall i)\ A_i \geqslant B_i, \quad (\exists j)\ A_j > B_j,$$

$$(\forall \omega \in \Omega)\ \|\{A_i : \omega \in A_i\}\| + 1 = \|\{B_i : \omega \in B_i\}\|.$$

See Kaplan (1974) for a proof although a more direct proof can be based upon the elementary properties of convex sets. In view of Theorem 3 we can establish

THEOREM 4. *If* $\geqslant$ *is strictly non-additive,* $\{A_i\}$ *are the sets referred to in Theorem 3, and* $\bar{s}$, $\underline{s}$ *are any upper and lower scales for* $\geqslant$, *then*

$$\sum_i \sum_{\omega \in A_i} [\bar{s}(\omega) - \underline{s}(\omega)] \geqslant \sum_{\omega \in \Omega} \underline{s}(\omega).$$

Hence, whenever $\underline{s}$ is non-trivial (not identically zero) we have that $\bar{s}$, $\underline{s}$ cannot agree too closely. This result can then be applied to establish

THEOREM 5. *If* $\geqslant_1$ *is strictly non-additive then there are infinitely many additive relations* $\geqslant_2$ *such that there is no complete ordering* $\geqslant$ *of* $\{(\Omega_1, A_2), (A_1, \Omega_2): A_i \in \mathfrak{A}\}$ *satisfying PCP1–PCP6 and*

$$(\Omega_1, A_2) \geqslant (\Omega_1, A_2') \Leftrightarrow A_2 \geqslant_2 A_2',$$
$$(A_1, \Omega_2) \geqslant (A_1', \Omega_2) \Leftrightarrow A_1 \geqslant_1 A_1'.$$

Hence we cannot compare the uncertainty of event A_1 from $\mathscr{E}_1$ with that of event A_2 from $\mathscr{E}_2$ for all $A_i \in \mathfrak{A}_i$ when our knowledge is that $\mathscr{E}_1$ is of non-additive type and $\mathscr{E}_2$ is any of infinitely many other random CP experiments.

Insofar as we do not insist upon the universal comparability of uncertainty and chance phenomena then we do not have access to a particular defense for our restriction to only quantitative models. In the absence of such a defense an attempt to use CP models that are not of additive type seems justified as it allows us more flexibility to fit real world phenomena.

7. THE QUANTIZATION OF UNCERTAINTY

A consequence of the lack of universal comparability of uncertainties is that there can be limits to the precision to which uncertainties about events in a given experiment can be measured. One form that such a limitation on precision can take is exhibited by the use of approximating probabilities $\underline{P}$, $\bar{P}$ to a CP order $\geqslant$. The additive set functions $\underline{P}$, $\bar{P}$ defined on an algebra $\mathfrak{A}$ are lower and upper approximating probabilities if:

$$(\forall A \in \mathfrak{A})\ 1 \geqslant \bar{P}(A) \geqslant \underline{P}(A) \geqslant 0;$$

$$A \geqslant B \text{ implies } P(A) \geqslant P(B).$$

When $\geqslant$ is additive or almost additive then we can find $\underline{P}$, $\bar{P}$ such that $\underline{P} \equiv \bar{P}$. However, if $\geqslant$ is strictly non-additive then it is an immediate consequence of Theorem 4 that there exists $\{A_i\}$ such that for any upper and lower approximating probabilities $\bar{P}$, $\underline{P}$ it follows that

$$\sum_i [\bar{P}(A_i) - \underline{P}(A_i)] \geqslant \underline{P}(\Omega).$$

Hence if we choose $\underline{P} \not\equiv 0$ then the upper and lower approximating probabilities cannot agree too closely.

It is no surprise to naive introspection that some uncertainties or chance phenomena may not admit of arbitrarily precise numerical description. That such precision is always possible is the burden of those who assert that the usual quantitative probability concept is universally applicable, and I am unaware of any arguments that establish this.

8. Frequentist interpretation of CP

The interpretations that have been associated with CP have been decision-oriented subjectivist (de Finetti, Savage), inference-oriented logical (Carnap, Keynes), and aesthetic (Koopman – not clear perhaps this is just subjectivistic), but never frequentist. We will propose a possible frequentist interpretation that can yield non-additive PCP but about which little is yet known. The usual frequentist interpretations of probability are based upon the possibility of an arbitrarily long, un-linked sequence of repetition of an experiment. Given a sequence $\mathbf{x} = x_1, ..., x_n, ...,$ of outcomes of the repeated experiment $\mathscr{E} = (\mathfrak{A}, P)$ or $(\mathfrak{A}, \geqslant)$, $x_i \in \Omega$, we compute the number of occurrences $N_A(x, n)$ of the event $A \subset \Omega$ in the first n outcomes $x_1, ..., x_n)$. The probability $P(A)$ of the event A is a measure of the tendency (propensity) for A to occur in a performance of $\mathscr{E}$ and is reflected in the relative frequency

$$N_A(\mathbf{x}, n)/n,$$
$$P(A) \approx N_A(n)/n.$$

The sense of approximation indicated by $\approx$ is of course problematic. Those who hold a limit view either assume that there is a Collective $\{x_i\}$ in which

$$\lim_{n \to \infty} N_A(n)/n \text{ exists and equals } P(A),$$

or there is an ensemble for which at least a weak law of large numbers (LLN) is valid and

$$(\forall \varepsilon > 0) \lim_{n \to \infty} P(|N_A(n)/n - P(A)| > \varepsilon) = 0.$$

As a 'practical' matter it is assumed that we can select n sufficiently large so that it is a 'moral certainty' that

$$|N_A(n)/n - P(A)| < \varepsilon.$$

The virtues and problems of a frequentist conception have been discussed at great length in many places and we content ourselves to refer to Fine (1970, 1973), Gillies (1973), Hacking (1965), and Salmon (1966).

One difficulty with a frequentist interpretation is that it seems to apply only to a sequence of outcomes that is 'statistical stable'. Yet we have no way of assuring ourselves that any given performance of unlinked repetitions of $\mathscr{E}$ will yield such outcomes. While our reliance on the apparent convergence of relative frequencies can be somewhat accounted for in various ways (Fine, 1970) it would be desirable to reduce our dependence on this requirement.

We propose to coordinate $\geq$ with frequencies of occurrence as follows. Let

$$\mu_{A,B}(\mathbf{x}, n) = \|\{j: N_A(\mathbf{x}, j) > N_B(\mathbf{x}, j), j \leq n\}\|,$$
$$\Delta_{A,B}(\mathbf{x}, n) = \mu_{A,B}(\mathbf{x}, n) - \mu_{B,A}(\mathbf{x}, n).$$

Introduce thresholds $\tau(\mathbf{x}, n) \geq \varepsilon(\mathbf{x}, n) \geq 0$. We suggest that

$$A > B \quad \text{if} \quad \Delta_{A,B} > \tau,$$
$$A \sim B \quad \text{if} \quad \text{MAX}[\mu_{A,B}, \mu_{B,A}] \leq \varepsilon.$$

Roughly speaking A is more probable than B if as we sequentially repeat $\mathscr{E}$ we find that most of the time there are more occurrences of A than of B. Alternatively, if we examine the outcome sequence $\mathbf{x}$ at a random time j then if $A > B$ the probability that we will find $N_A(\mathbf{x}, j) > N_B(\mathbf{x}, j)$ exceeds that for j such that $N_R > N_A$. Two events are deemed equally probable if neither of N_A, N_B, leads the other for an appreciable part of the repeated experiment.

Note that the suggested interpretation does make use of all of the data $\mathbf{x}$ and thus does not contradict our intuition that the most reliable assessment of the tendency for events to occur should be based upon all of the data; albeit it does contradict the usual conclusion that only the last values of $N_A(\mathbf{x}, j)$ are relevant.

A tedious example has confirmed that there are outcome sequences $\mathbf{x}$ from which the suggested interpretation generates a complete CP relation of non-additive type. It is unclear whether all CP relations can be generated as above from suitable occurrence sequences.

If we are interested in a limit interpretation then we might interpret $A > B$ by $\Delta_{A,B}(x, n) < \tau$ only finitely often in n. If in fact

$$\lim_{n \to \infty} N_A/n = P(A), \qquad \lim_{n \to \infty} N_B/n = P(B),$$

then $A > B$ if $P(A) > P(B)$ provided that

$$\limsup_n \tau/n < 1,$$

and this agrees with the usual frequency interpretation. Of course $\Delta_{A,B} < \tau$ only finitely often may occur even when no relative frequencies converge and even when N_A and N_B change lead infinitely often. Hence our frequentist interpretation broadens the applicability of probability by supplying a probabilistic description for many sequences of outcomes that had hitherto been ruled out of consideration. The requirement of statistical stability has been substantially weakened, albeit we still face the similar inductive problem of inferring from a finite sequence of outcomes to the unobservable contention that $\Delta_{A,B} < \tau$ only finitely often.

One word of caution is in order concerning our suggested interpretation. If τ is chosen too small and/or ε too large then the resulting PCP relation may violate transitivity. Formally we have the easily proven

THEOREM 6. *If $\varepsilon < n/3$, $n \geqslant \tau \geqslant \frac{2}{3}n$, then $A > B$, $A \sim B$ according to whether $\Delta_{A,B}(\mathbf{x}, n) > \tau$ or $\mathrm{MAX}[\mu_{A,B}, \mu_{B,A}] \leqslant \varepsilon$ will satisfy PCP1–PCP6 for all $\mathbf{x}$ and n.*

Reducing τ below $2n/3$ or increasing ε above $n/3$ can yield violations of PCP3 or PCP6 depending upon $\mathbf{x}$.

Our suggested frequentist interpretation of CP or PCP enables us to substantially increase the domain of applicability of probability in a manner substantially (completely, if we refer only to limit interpretations) in agreement with the usual frequentist interpretation of quantitative probability. We take the broader applicability of CP as a significant argument in favor of its use.

Just as there are many different statistically motivated actual ways to estimate quantitative probability from frequency of occurrence data, so should it be possible to introduce variations of the relation between the PCP relation $\geqslant$ and frequency data. We have not yet examined these possibilities.

A different analysis of a frequentist interpretation that could yield non-additive CP relations conforming to frequency of occurrence data can be developed based upon the possibility that in some domains of applications (e.g. quantum mechanics)

(i)   Counting errors in the determination of N_A may be intrinsically unavoidable;

(ii) Comparison of the relative tendencies for events A and B to occur may necessitate a different experiment than would suffice for events C and D, and the results of all pairwise comparisons might then assemble into a non-additive PCP relation.

We will not pursue this inquiry here.

9. TOWARDS A LAW OF LARGE NUMBERS

We believe that the laws of large numbers (LLN) support a frequency interpretation only by establishing a sort of self-consistency for the interpretation. The LLN do not guarantee the success of a frequency interpretation. However, they do establish the existence of circumstances under which the frequency interpretation of probability is sensible. It appears likely that a weakened version of the LLN can be established to defend our proposed frequency interpretation of CP. There is of course the obstacle, reported in Theorem 1, that we can only consider complete extension of a CP experiment to arbitrarily many IID experiments when it is of additive type.

While we have not yet established a CP version of a LLN we believe it should be possible to do so along the following lines. To model an arbitrarily long sequence of unlinked repetitions of $\mathcal{E} = (\mathfrak{A}, \geqslant)$ we extend $\mathcal{E}$ to $\mathcal{E}^{(n)} = (\mathfrak{A}^{(n)}, \geqslant^{(n)})$ where $\geqslant^{(n)}$ is now a maximal PCP relation between elements of $\mathfrak{A}^{(n)}$ satisfying PCP1–PCP6 and the requirement that it be of IID type as defined in Section 4. By $\geqslant^{(n)}$ being maximal we mean that if $A^{(n)}$, $B^{(n)} \in \mathfrak{A}^{(n)}$ are not compared by $\geqslant^{(n)}$, then extending $\geqslant^{(n)}$ to include any of $A^{(n)} > B^{(n)}$, $A^{(n)} \sim B^{(n)}$, $B^{(n)} > A^{(n)}$, will yield a contradiction of either some of PCP1–PCP6 or the properties of an IID type extension. When $\geqslant$ is non-additive then it follows from Theorem 1 that for sufficiently large n, $\geqslant^{(n)}$ cannot be a complete ordering of events.

A weak but useful law of large numbers could be established if we could prove that for all $\geqslant$ satisfying PCP1–PCP6 and for all sufficiently large n, there exists $\geqslant^{(n)}$ of IID type such that if

$$R_{A,B}(n) = \{(x_1, ..., x_n): \Delta_{A,B} > \tau \text{ or } \mathrm{MAX}\,[\mu_{A,B},\, \mu_{B,A}] \leqslant \varepsilon\}$$

then

$$R_{\geqslant}(n) \equiv \bigcap_{\{(A,B):\, A \geqslant B\}} R_{A,B}(n) >^{(n)} \overline{R_{\geqslant}(n)}$$

Such a LLN could then be refined to determine the extent to which $R_{\geq}(n)$ is more probable than its complement. It is not apparent to us that as n increases, $R_{\geq}(n)$ will necessarily converge in certainty to the certainty of Ω. If $R_{\geq}(n)$ does not for large enough n become at least as probable as any fixed event $A^{(m)} \in \mathfrak{A}^{(m)}$ where $\bar{A}^{(m)} > \varphi$, then one of the questions about the LLN raised and answered by Bernoulli (1713) will be reopened.

Cornell University, Ithaca

BIBLIOGRAPHY

Bernoulli, J.: 1713 (Translated 1966), *Ars Conjectandt-part IV*, Dept. of Statistics, Harvard University, Cambridge, Mass.
Bernstein, S. N.: 1917 (Reprinted 1964), 'Axiomatic Foundations of Probability Theory', *Collected Works IV*, Academy of Science, USSR, 10–25.
Carnap, R.: 1950 (2nd Ed. 1962), *Logical Foundations of Probability*, Univ. of Chicago Press, Chicago, pp. 428–462.
Domotor, Z.: 1969, *Probabilistic Relational Structures and Their Applications*, Tech. Rep. 144, Inst. for Math. Studies in Social Sciences, Stanford University, Stanford.
Fine, T. L.: 1970, 'On the Apparent Convergence of Relative Frequency and Its Implications', *IEEE Trans. on Information Theory* IT-16, 251–257.
Fine, T. L.: 1971a, 'A Note on the Existence of Quantitative Probability', *Ann. Math. Stat.* **42**, 1182–1186.
Fine, T. L.: 1971b, 'Rational Decision Making with Comparative Probability', *Proc. IEEE Conf. on Decision and Control Miami Beach 1971*, Dept. of Elec. Eng. Univ. of Florida, Gainesville, pp. 355–356.
Fine, T. L.: 1973, *Theories of Probability*, Academic Press, New York, pp. 15–57.
Fine, T. L.: 1974, 'Towards a Revised Probabilistic Basis for Quantum Mechanics,' *Synthese* **29**, 187–201.
Fine, T. L. and J. T. Gill: 1976, 'The Enumeration of Comparative Probability Relations', *Ann. Probability*, to appear.
Fine, T. L. and M. A. Kaplan: 1976, 'Joint Orders in Comparative Probability', to appear *Ann. Probability*.
Finetti, B. de: 1931, 'Sul Significato Soggettivo della Probabilita', *Fundamenta Mathematicae* **31**, 298–329.
Fishburn, P. C.: 1970, *Utility Theory for Decision Making*, J. Wiley & Sons, New York, pp. 193–201.
Fishburn, P. C.: 1975 'Weak Comparative Probability on Infinite Sets', *Ann. of Probability* **3**, 889–893.
Gillies, D.: 1973, *An Objective Theory of Probability*, Methuen, London.
Hacking, I.: 1965, *Logic of Statistical Inference*, Cambridge Univ. Press, London and New York, 1–53.

Kaplan, M. A.: 1971, *Independence in Comparative Probability*, Cornell University, Ithaca, N.Y., M.S. thesis.

Kaplan, M. A.: 1974, *Extensions and Limits of Comparative Probability Orders*, Cornell University, Ithaca, N.Y. Ph.D. thesis.

Keynes, J. M.: 1921 (Reprinted 1962), *A Treatise on Probability*, Harper and Row Publishers, Inc., New York.

Koopman, B. O.: 1940, 'The Bases of Probability', *Bull, Amer. Math. Soc.* **46,** 763–774.

Kraft, C., Pratt, J., and Seidenberg, A.: 1959, 'Intuitive Probability on Finite Sets', *Ann. Math. Stat.* **30,** 408–419.

Krantz, D., R. Luce, P. Suppes, and A. Tversky: 1971, *Foundations of Measurement*, Vol. I, Academic Press, New York, 199–244.

Luce, R. D.: 1967, 'Sufficient Conditions for the Existence of a Finitely Additive Probability Measure', *Ann. Math. Stat.* **38,** 780–786.

Luce, R. D.: 1968, 'On The Numerical Representation of Quantitative Conditional Probability', *Ann. Math. Stat.* **39,** 481–491.

Narens, L.: 1974, 'Minimal Conditions for Additive Conjoint Measurement and Quantitative Probability', *J. Math. Psychology.* **11.**

Roberts, F. S.: 1973, 'A Note on Fine's Axioms for Quantitative Probability', *Ann. Probability*, 484–487.

Salmon, W.: 1966, *The Foundations of Scientific Inference*, Univ. of Pittsburgh Press, Pittsburgh.

Savage, L. J.: 1954, *The Foundations of Statistics*, J. Wiley & Sons, New York, pp. 34–36.

Scott, D.: 1964, 'Measurement Structures and Linear Inequalities', *J. Math. Psychology* **1,** 233–247.

Villegas, C.: 1964, 'On Qualitative Probability σ-Algebras', *Ann. Math. Stat.* **35,** 1787–1796.

ILKKA NIINILUOTO

ON THE TRUTHLIKENESS OF GENERALIZATIONS

1. THE PROBLEMS OF TRUTHLIKENESS

The topic of this paper is the notion of truthlikeness. First, the logical problem of truthlikeness is considered: Popper's definition of verisimilitude is criticized in Sections 2 and 3, and an alternative approach is outlined within a simple and precise framework in Sections 4 and 5. The epistemic problem concerning methods of estimating degrees of truthlikeness on the basis of some evidence is discussed in Section 6. In the final Section 7, the methodological import of this notion is briefly commented on.

Truthlikeness as a methodological and philosophical idea has a very interesting history. In the middle of the 17th century, the physicists Robert Boyle and Robert Hooke compared the method of science to the rule called *regula falsi* in arithmetic. *Regula falsi* is a method of solving a linear equation by proceeding from a false guess of the solution; the solution sought for is obtained with the help of one conjecturally supposed number. In a similar way, Boyle argued, it may be useful in the search of truth in science to proceed from false assumptions or hypotheses. Referring to Francis Bacon, he remarked that "it has truly been observed by a great philosopher, that truth does more easily emerge out of error than confusion".[1] This assumed analogy between science and mathematics was extended, in the 18th century, to the iterative methods of approaching the solutions of arithmetical equations by a process of successive approximation. David Hartley, Joseph Priestley, and Georges Le Sage – early defenders of the hypothetico-deductive method in science – argued that, in a similar way, all advance in science is carried on by the use of self-corrective methods which gradually eliminate falsehood and eventually bring the scientific theories to a closer and closer approximation to the truth.[2] In the 19th century, this view about the scientific method was also

Butts and Hintikka (eds.), Basic Problems in Methodology and Linguistics, 121–147.
Copyright © 1977 by D. Reidel Publishing Company, Dordrecht-Holland. All Rights Reserved.

influenced by the dynamic theory of human knowledge that German idealists had developed.[3] As a result, the idea of 'approach to the truth' is a crucial ingredient of the methodological views of such different philosophers as William Whewell, Charles Peirce, Pierre Duhem, Friedrich Engels, V. I. Lenin, Hans Reichenbach and Karl Popper.

The idea that science follows a method which is hoped to lead to the truth at least in the long run is important for the 'realist' theory of scientific progress. In contrast with the methodological instrumentalists, the so called scientific realists regard scientific theories as attempts to express *true information about the reality*. In this view, which I am willing to share, the primary aim of science is informative truth, and an essential aspect of the progress in science is to be defined in terms of this aim.[4] Even though we may admit that most of our theories are idealized and imprecise, simplified and perhaps false, still they can be regarded as stepping-stones to better representations of the reality.[5] To defend this position, the realists have to make sense of such notions as 'approximate truth', 'closeness to the truth', and 'information about the reality'. This paper is hoped to clarify some aspects of this task which – as it has turned out – faces some surprising obstacles.

Much of the recent discussion about the notions of approximate truth and truthlikeness has been stimulated or provoked by Sir Karl Popper's theory of *verisimilitude*, which in turn was inspired by Quine's critical remarks on Peirce's view of truth as the limit of inquiry.[6] Let T and F be the sets of all the true and false sentences about the actual world, respectively, and let S and S' be two successive scientific theories (i.e. deductively closed sets of sentences). Then a naive cumulative view of scientific progress assumes that S and S' are related to each other in the following way: $S \subseteq S' \subseteq T$. In other words, theory S is a collection of true sentences, and its successor S' comprises a larger share of the set T of true sentences. Against this view, Popper recognizes that scientific theories are usually, in the strict sense, false – but false theories nevertheless have true consequences. The *truth content* of theory S is defined by Popper as the intersection $S \cap T$ of S and T, while the *falsity content* of S is the intersection $S \cap F$ of S and F. Popper's qualitative definition of truthlikeness is the following: theory S' is closer to the truth than theory S if and only if $S \cap T \subseteq$

$S' \cap T$ and $S' \cap F \subseteq S \cap F$, where at least one of these containments is strict.

It has been shown recently by David Miller, Pavel Tichý, and John Harris that this definition is "on the wrong track altogether".[7] Two false theories cannot stand in the above defined relation: if S' is closer to the truth than S, then S' has to be true. Moreover, if T is unaxiomatizable and S' is axiomatizable, then S' is closer to the truth than S only if both S and S' are true. There seems to be no escape from these results, which even can be generalized.[8] Popper's qualitative notion of verisimilitude – and for similar reasons his quantitative measure of verisimilitude as well – is thus entirely inadequate for the purpose it was designed, viz. making sense of the comparison of the truthlikeness of two false theories.

David Miller has also pointed out difficulties in another attempt to characterize verisimilitude, which he attributes to John Watkins. Miller shows that "no false quantitative theory is uniformly more accurate than is any other" (Miller, 1975a, p. 159). Miller does not find anything positive to sweeten his sceptical conclusions: as there are no prospects of making sense of the notion of truthlikeness, he argues, there does not seem to be any objective reasons for preferring one false theory to another for theoretical or practical purposes.

In this paper, I wish to add to these interesting negative results something of a more constructive nature. The notion of truthlikeness is here discussed in a special case which clearly does not cover all the situations where this notion is usually thought to be applicable.[9] But as my aim is not that of devising measures of truthlikeness for the use of working scientists, but rather a philosophical analysis of a notion which has been treated with great suspect by many philosophers, the use of a simple framework seems here well motivated. Moreover, as the underlying idea of this paper is quite general, it seems evident that at least some of the general characteristics of truthlikeness will be captured in my approach.

The approach of this paper continues and modifies the possible worlds account of truthlikeness that Risto Hilpinen presented in 1974 at the Warsaw Conference of Formal Methodology (see Hilpinen, 1976). He showed that the notion of truthlikeness presupposes a similarity metric – that is, a quantitative 'distance measure' or at least a comparative counterpart to such a measure – on the set of possible worlds. This general idea is adapted in this paper to the treatment of the truthlikeness of generalizations.

2. THE WHOLE TRUTH

Popper assumes that "if we speak about approach or approximation to truth, we mean to 'the whole truth'; that is, the whole class of true statements, the class T" (Popper, 1972, p. 55). In this sense, a theory has the maximum truthlikeness just in case it is not only true, i.e. has no false consequences, but also implies all the true statements. This remark of Popper's serves well to remind that his notion of verisimilitude is something else than 'degree of truth' in the sense of 'degree of truth value'; it is rather an attempt to combine the ideas of truth and content. It seems, however, that for at least two different reasons the proposal of using the class T as the basic ingredient of the definition of truthlikeness is not a fortunate one.

In the first place, all the true statements about the actual world, or about any given domain of objects, do not constitute a well-defined class. Talk about the class T presupposes that the underlying language or vocabulary L has been fixed. But it is highly questionable whether any single language L can be used to describe the reality in all its variety. At least no scientific theory attempts to do this: theories have restricted vocabularies and they try to describe only those aspects of the reality that are expressible in their own language. On the other hand, if science does progress towards the truth, this process certainly involves conceptual change or conceptual enrichment. To do justice to this dynamic feature of science, we should find some reasonable way of making the notion of truthlikeness relative to the expressive power of theories.

Secondly, even though one may meaningfully talk about the class of all the true sentences of a fixed language L, this class contains many sentences which the scientists are not particularly interested in. The proper subject matter of science includes only those aspects of the world which are, in some sense, uniform or regular. With some qualifications, we may say that genuine scientific knowledge does not contain singular statements about particular individuals, but is expressible in universal form. To the extent that this claim is valid, it is more appropriate to consider the set of all the *true generalizations expressible in L* than the set of all the true sentences in L.

To illustrate these remarks, let L be a monadic first-order language with the set $\lambda = \{O_1, ..., O_k\}$ of logically independent primitive predicates, with Ct-predicates Ct_j, $j = 1, ..., K$ (here $K = 2^k$), and with constituents C_i, $i = 1, ..., 2^K$.[10] Then the set G of all generalizations in L can be defined as the disjunctive closure $D(B)$ of the set B of all constituents of L:

$$(1) \qquad G = D(B) = \left\{ \bigvee_{i \in J} C_i \mid \varnothing \neq J \subseteq \{1, 2, ..., 2^K\} \right\}.$$

Thus, each generalization h in G can be expressed in an L-*normal form*, that is, for each h in G there is a set $\mathbf{I}_h \subseteq \{1, 2, ..., 2^K\}$ such that

$$(2) \qquad \vdash h \equiv \bigvee_{i \in \mathbf{I}_h} C_i.$$

Let L_1 and L_2 be two monadic first-order languages with the same individual constants and with the finite sets λ_1 and λ_2 of primitive one-place predicates, respectively. If L_2 is an extension of L_1, i.e. $\lambda_1 \subseteq \lambda_2$, then each L_1-constituent in B_1 is logically equivalent to a finite disjunction of L_2-constituents in B_2. This relation is denoted by $B_1 \preccurlyeq B_2$, and the set B_2 of L_2-constituents is said to be at least as *fine* as the set B_1 of L_1-constituents. Thus,

$$(3) \qquad B_1 \preccurlyeq B_2 \quad \text{iff} \quad B_1 \subseteq D(B_2) \quad \text{iff} \quad G_1 \subseteq G_2.$$

Therefore, each generalization h in L_1 can be expressed in an L_2-normal form with respect to any extension L_2 of L_1. Moreover, at least if the new predicates in $\lambda_2 - \lambda_1$ are not explicitly definable in terms of λ_1, then language L_2 gives us the possibility of making a finer description of the reality. In particular, the true constituent C_t^2 of L_2 entails the true constituent C_t^1 of L_1, but not conversely, so that C_t^2 has a greater logical content than C_t^1.

More generally, if L_1, L_2, ..., L_n, ... is a sequence of monadic first-order languages such that

$$\lambda_1 \subseteq \lambda_2 \subseteq \ldots \subseteq \lambda_n \subseteq \ldots$$

then for the corresponding sets of constituents

(4) $B_1 \leqslant B_2 \leqslant \ldots \leqslant B_n \leqslant \ldots$

and for the corresponding sets of generalizations

(5) $G_1 \subseteq G_2 \subseteq \ldots \subseteq G_n \subseteq \ldots$

The corresponding true constituents C_t^1, C_t^2, ..., C_t^n, ... constitute a sequence of generalizations with increasing logical strength.

Let h be a generalization in a language L which is interpreted on a domain U. Then the true constituent C_t gives a complete description of U in terms of quantificational sentences employing the extra-logical vocabulary λ of L. What generalization h expresses about the domain U can now be compared to the truth about U as expressed by C_t. In other words, we may ask, how close h is to the true constituent C_t of L. However, language L may be the starting point of an infinite number of infinite sequences of languages L_i, $i = 1, 2, \ldots$, of the type (4). As h is logically equivalent to a generalization in each language L_i which is an extension of L, we may also ask, how close h is to the true constituent C_t^i of L_i. But it does not seem to make sense to ask, how close h is to the truth – if by 'the truth' is meant something like 'the whole truth' about U. As there are innumerable finer true constituents which entail the true constituent C_t of L, this question seems to make as little sense as the question, how close the natural number 100 is to the infinity.

Instead of comparing the given generalization h to 'the whole truth', or even to the set of all the true sentences in some language L_i, generalization h may be compared to the true constituent C_t^i of language L_i, where C_t^i gives a structural description of the domain U and entails all the true generalizations of L_i. In other words, the notion of truthlikeness presupposes some conceptual system which indicates the *depth of analysis*, as we might say, at which we are considering the given generalization. A measure of truthlikeness does not reflect the 'distance' of h from 'the whole truth', but rather from the most informative true generalization at some given depth of analysis.

3.   LINGUISTIC INVARIANCE AND TRUTHLIKENESS

Miller has suggested that perhaps it is not necessary to take "the fluidity of scientific language into account when discussing truthlikeness" (Miller, 1975b, p. 207). Can we not, he asks, always combine the language of two theories into a language adequate for them both? Moreover, truthlikeness should be insensitive to simple linguistic reformulation – for Miller, "truthlikeness, like truth itself, has no appeal if not invariant under translation" (*ibid.*, pp. 214–215).

Miller seems here to overlook the fact that, as truthlikeness is a combination of truth *and* content, we cannot simply expect it to be analogous to truth in all important respects. Perhaps linguistic invariance should be required of the truth-factor of truthlikeness, but it is argued below that it cannot be required of the information-factor. The amount of information about the truth that a generalization h contains depends essentially upon the depth of analysis at which we are considering h. An example illustrating this claim is given in Figure 1.

Figure 1 shows how classification systems become finer and finer by means of the introduction of new predicates. At the first level, L_1, there is only one cell, and it may be either empty (white square) or instantiated (black square). When a new monadic predicate M_1 is

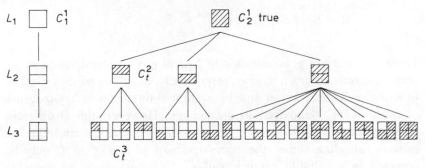

Fig. 1.

introduced into the language, each cell Ct_j of L_1 splits into two subcells, defined by Ct_j & M_1 and Ct_j & $\sim M_1$. These subcells are then again split into two subcells by means of a new predicate M_2. At the first level, L_1, there are two different constituents, at the second level, L_2, there are four constituents, and at the third level, L_3, there are sixteen constituents. All subcells of an empty cell remain empty, no matter how deep we go in the analysis. But almost anything can happen to the subcells of a non-empty cell – as long as one of them remains non-empty. In Figure 1, the true constituent C_2^1 of L_1 splits, at the third level L_3, into a true but uninformative disjunction of *fifteen* L_3-constituents – many of these disjuncts are, intuitively speaking, very far from the true L_3-constituent C_t^3. On the other hand, the false constituent C_1^1 at the level L_1 is equivalent to *one* L_3-constituent – and this constituent is not very far from the true one.[12]

This example suggests that the relative distances of constituents from the true one need not be preserved when greater depths of analysis are considered. More generally, let h and g be two generalizations in a language L_1, and let h' and g' be their 'translations' into a language L_2, where L_2 is an extension of L_1. Then, even if h is closer to the truth than g at the level of L_1, g' may be closer to the truth than h' at the level of L_2.

This observation shows also why Miller's suggestion of combining the language of two theories does not generally work. Let L_1 and L_2 be two languages such that neither $\lambda_1 \subseteq \lambda_2$ nor $\lambda_2 \subseteq \lambda_1$, and let L be the mutual extension of L_1 and L_2, that is, a language with the vocabulary $\lambda = \lambda_1 \cup \lambda_2$. For the corresponding sets of constituents, we have then

$$B_1 \leqslant B \quad \text{and} \quad B_2 \leqslant B.$$

Therefore, if h is a generalization in L_1 and g is a generalization in L_2, they have translations h' and g', respectively, in L. The truthlikeness of h' and g' *relative to* L can now be evaluated by means of any measure of truthlikeness relativized to a language. However, this is different from comparing the original generalizations h and g for truthlikeness in some absolute sense: the comparison of h' and g' in L may be reversed in a suitable extension of L. In other words, in order to

compare two generalizations in two different languages for their truthlikeness, we first have to *make them comparable* by translating them
into a suitably chosen new language. This translation may, however,
involve radical conceptual change which is not reducible to simple
linguistic reformulation, so that the content of the original generalizations may change as well.[13]

4. DISTANCE BETWEEN CONSTITUENTS

It has been argued in Sections 2 and 3 that a reasonable notion of
truthlikeness should be relative to some conceptual system or some
vocabulary. The reason for this 'conceptual relativity' of truthlikeness
was seen to be the fact that some generalizations contain 'deeper'
information than others – information which becomes relevant through
conceptual enrichment.

The notion of truthlikeness is not only relative to a conceptual
system L, but also to some distance measure between constituents of L
and to some way of extending this measure to all generalizations of L.
In Stalnaker's terms,[14] we may say that there are proposed below
several logical notions of 'truthlikeness relative to L' which are *semantically unambiguous* but *pragmatically ambiguous*. The situation here is
similar (even if not complete analogous) to the case of different maps
describing the network of roads and railways in the United States. If
map S_1 shows all the motorways and main highways, while map S_2
shows all the motorways, main highways, main roads, and some
secondary roads as well, then S_2 as a more complete description is
'closer to the truth' than S_1. But if map S_3 shows all the motorways
and railways, is it 'closer to the truth' than S_1? Is S_2 still 'closer to the
truth' than S_1 if it makes mistakes about the secondary roads? The
answers to these questions presuppose that some relative weights are
given to the comparative relevance of the main highways against
railways or of the main roads against secondary roads. In a similar way,
the pragmatic ambiguity of truthlikeness means that all the claims
about the comparative truthlikeness of generalizations presuppose the
specification of a distance measure which reflects the relevant respects
of comparison and balances them against each other.

The problem of defining the distance between monadic constituents has a simple and straightforward solution.[15] Constituents of a language L with k primitive predicates can be compared only in $K = 2^k$ respects, viz. the K cells defined by the Ct-predicates Ct_j of L. Each of these respects of comparison admits two values: same or different. The distance between two L-constituents can thus be defined as the number of cells relative to which they make different claims. This definition can be formalized in the following way. There is a natural one-to-one correspondence between L-constituents and binary sequences of length K: the sequence $x = x_1 x_2, ..., x_K$ corresponds to constituent C_j if and only if for all $i = 1, ..., K$

$$x_i = 1, \quad \text{if} \quad Ct_i \in \mathbf{CT}_j$$
$$= 0, \quad \text{otherwise.}$$

Then the *distance* $d(x, y)$ between constituents x and y can be defined by

$$(6) \qquad d(x, y) = \sum_{i=1}^{K} |x_i - y_i|.$$

A more general measure, d', could be defined by

$$(7) \qquad d'(x, y) = \sum_{i=1}^{K} a_i |x_i - y_i|,$$

where $a_i \geqslant 0$ for $i = 1, ..., K$. In formula (7), the relative importance of the ith cell Ct_i is indicated by the non-negative constant a_i. Measure d is a special case of measure d', where $a_i = 1$ for all $i = 1, ..., K$. For simplicity, a principle of equality among the cells Ct_i is assumed below, so that only the measure d is considered in this paper.

According to definition (6), we have $0 \leqslant d(x, y) \leqslant K$ for all L-constituents x and y. In particular, $d(x, y) = 0$ if and only if $x = y$. Moreover, the number of different constituents at the distance m from constituent x is $\binom{K}{m}$. For constituents x, y, and z, define

$$(8) \qquad x <_y z \quad \text{iff} \quad d(x, y) < d(z, y)$$

(9) $x \leqslant_y z$ iff $d(x, y) \leqslant d(z, y)$

(10) $x \approx_y z$ iff $d(x, y) = d(z, y)$.

Then $x <_y z$ iff x is *closer* to y than z; $x \leqslant_y z$ iff z is not closer to y than x; and $x \approx_y z$ iff x and z are *equally close* to y. For each L-constituent y, relation $\leqslant_y$ is a weak ordering of the set of L-constituents, and relation $\approx_y$ is an equivalence relation in this set. If for a constituent y and for $0 \leqslant i \leqslant K$ we define

(11) $N_i(y) = \{x \mid d(x, y) \leqslant i\}$,

then

(12) $\mathbf{N}(y) = \{N_i(y) \mid i = 0, ..., K\}$

is a *nested system of spheres* in the sense of Lewis (1973).

Example. Let $k = 2$, and define $Ct_1 = O_1 \& O_2$, $Ct_2 = O_1 \& \sim O_2$, $Ct_3 = \sim O_1 \& O_2$, and $Ct_4 = \sim O_1 \& \sim O_2$. Then the spheres $N_i(x)$ around the constituent $x = 1011$ are illustrated in Figure 2. The dotted line in Figure 2 surrounds those constituents which belong to the normal form of the generalization $(x)(O_1(x) \supset O_2(x))$.

5.   DEGREES OF TRUTHLIKENESS

To define degrees of truthlikeness in the way outlined in Section 2, the distance measure d has to be extended (in the first argument) to arbitrary generalizations of language L. In virtue of the normal form (2), there is a one-to-one correspondence between the consistent generalizations of L and the non-empty subsets of the set B of L-constituents. (Cf. Figure 2.) The problem of defining degrees of truthlikeness reduces thus to the problem of evaluating the distance of a set of constituents from a single constituent, viz. from the true constituent C_t.

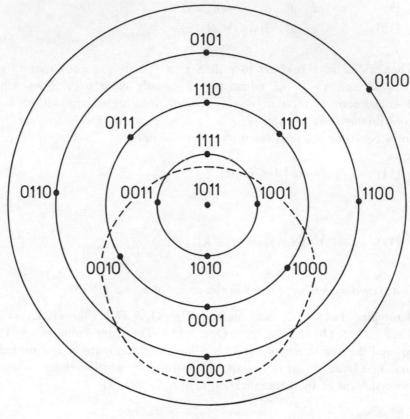

Fig. 2.

Let h be any consistent generalization of language L. Define functions m_* and m^* as follows:

(13) $m_*(h, C_t) = \min_{h \in \mathbf{I}_h} d(C_i, C_t)$

(14) $m^*(h, C_t) = \max_{h \in \mathbf{I}_h} d(C_i, C_t)$.

Then $m_*(h, C_t)$ gives the minimum distance of h from C_t, while $m^*(h, C_t)$ gives the corresponding maximum distance. Thus, function m_* tells how close to C_t generalization h 'comes' at best, while function m^* tells how far from C_t it 'remains' at worst. If $\vdash h \equiv C_i$, then $m_*(h, C_t) = m^*(h, C_t) = d(C_i, C_t)$.

Assuming that C_t is the true L-constituent, define functions V and I as follows:

$$(15) \qquad V(h, C_t) = 1 - \frac{m_*(h, C_t)}{K}$$

$$(16) \qquad I(h, C_t) = 1 - \frac{m^*(h, C_t)}{K}.$$

Then function V has the following properties: for consistent generalizations h, h_1, and h_2 in L,

(V1) $0 \leqslant V(h, C_t) \leqslant 1$.

(V2) $V(h, C_t) = 1$ iff h is true.

(V3) $V(h, C_t) = 1$ or $V(\sim h, C_t) = 1$, but not both.

(V4) $V(h_1 \vee h_2, C_t) = \max\{V(h_1, C_t), V(h_2, C_t)\}$.

(V5) If h_1 & h_2 is consistent, then
$V(h_1, C_t) \geqslant V(h_1 \text{ \& } h_2, C_t)$.[16]

In view of these results, function V serves as a measure of the *degree of truth* – or degree of truth value – of generalizations. It gives the maximum value one to all the true generalizations in L – to logical truths as well as to informative ones.

According to definitions (14) and (16), function I has the following properties: for consistent generalizations h, h_1, and h_2 in L,

(I1) $0 \leqslant I(h, C_t) \leqslant 1$.

(I2) If h is logically true, then $I(h, C_t) = 0$.

(I3) $I(h, C_t) = 0$ or $I(\sim h, C_t) = 0$.

(I4) $I(h_1 \vee h_2, C_t) = \min\{I(h_1, C_t), I(h_2, C_t)\}$.

(I5) If h_1 & h_2 is consistent, then
$I(h_1, C_t) \leqslant I(h_1 \text{ \& } h_2, C_t)$.[17]

Function I takes a high value, if generalization h does not allow situations which are far from the truth as described by C_t. To use

Hilpinen's phrase, we may say that function I measures the *information about the truth* in generalization h (at the depth of C_t).

For L-constituents C_i, we have

$$(17) \qquad V(C_i, C_t) = I(C_i, C_t) = 1 - \frac{d(C_i, C_t)}{K}.$$

In general, a generalization may have a high degree of truth but a low degree of information about the truth – e.g. logical truths. But as

$$(18) \qquad V(h, C_t) \geqslant I(h, C_t)$$

for all generalizations h in L, a generalization with a high degree of information will also have a high degree of truth. Function I is, however, insensitive to truth values in the following sense:

$$(19) \qquad \text{If } h \text{ is a false generalization in } L, \text{ then } V(h \vee C_t, C_t) = 1 \text{ and } I(h \vee C_t, C_t) = I(h, C_t).$$

On the other hand, maximal information about the truth C_t can be given only by C_t itself:

$$(20) \qquad I(h, C_t) = 1 \quad \text{iff} \quad \vdash h \equiv C_t.$$

If the notion of truthlikeness should contain both the ideas of truth and content, the above results show clearly that neither V nor I alone can be taken as measure of truthlikeness (cf. results (V2) and (19), especially). A definition of truthlikeness has to give some weight to the 'truth-factor' V and 'information-factor' I.[18] A measure M_1 of the *degree of truthlikeness* of a generalization h (at the depth of C_t) may be defined as follows:

$$(21) \qquad M_1(h, C_t) = \gamma V(h, C_t) + (1 - \gamma)I(h, C_t),$$

where the constant $\gamma \in [0, 1]$ is an index of the relative weight of V as compared to I. A special case of (21) with $\gamma = \frac{1}{2}$ gives

$$(22) \qquad M_2(h, C_t) = \frac{1}{2}(V(h, C_t) + I(h, C_t)).$$

M_2 is then simply the arithmetical mean of V and I.

Measures M_1 and M_2 take into account only the minimum and the maximum distances of h from C_t: they do not depend upon the number of different constituents in the normal form of h at different distances from C_t. Therefore, if C_i is a L-constituent with $d(C_i, C_t) \leqslant m^*(h, C_t)$, then $M_1(h \vee C_i, C_t) \geqslant M_1(h, C_t)$ and $M_2(h \vee C_i, C_t) \geqslant M_2(h, C_t)$. At least when $m^*(h, C_t)$ is high, this may seem unsatisfactory, since the addition of C_i to h may bring into consideration situations with a great distance from the truth C_t. To avoid this difficulty, truthlikeness may be defined in terms of the *mean distance* of h from C_t:[19]

$$(23) \qquad m(h, C_t) = \frac{1}{|\mathbf{I}_h|} \sum_{i \in \mathbf{I}_h} d(C_i, C_t).$$

Thus, a third alternative for defining the degree of truthlikeness of h relative to C_t is the following:

$$(24) \qquad M_3(h, C_t) = 1 - \frac{m(h, C_t)}{K}.$$

Measures M_1, M_2, and M_3 have the following properties: for consistent generalizations h, h_1, and h_2 in L,

(T1) $0 \leqslant M_i(h, C_t) \leqslant 1$, for all $i = 1, 2, 3$.

(T2) $M_i(h, C_t) = 1$ iff $\vdash h \equiv C_t$, for all $i = 1, 2, 3$.

(T3) $M_i(h, C_t) = 0$ iff $\vdash h \equiv C_j$, where $d(C_j, C_t) = K$, for all $i = 1, 2, 3$.

(T4) If h is logically true, then $M_1(h, C_t) = \gamma$ and $M_2(h, C_t) = M_3(h, C_t) = \frac{1}{2}$.

(T5) If h is true, then $M_1(h, C_t) = \gamma + (1 - \gamma)I(h, C_t)$ and $M_2(h, C_t) = \frac{1}{2}(1 + I(h, C_t))$.

(T6) When C_j is an L-constituent,

$$M_i(C_j, C_t) = 1 - \frac{d(C_j, C_t)}{K}, \quad \text{for all } i = 1, 2, 3.$$

$$(T7) \quad M_1(h_1 \lor h_2, C_t) = \gamma \max \{V(h_1, C_t), V(h_2, C_t)\}$$
$$+ (1 - \gamma) \min \{I(h_1, C_t), I(h_2, C_t)\}$$
$$M_2(h_1 \lor h_2, C_t) = \tfrac{1}{2}[\max \{V(h_1, C_t), V(h_2, C_t)\}$$
$$+ \min \{I(h_1, C_t), I(h_2, C_t)\}]$$
$$M_3(h_1 \lor h_2, C_t) = \frac{|\mathbf{I}_{h_1}| \, M_3(h_1, C_t) + |\mathbf{I}_{h_2}| \, M_3(h_2, C_t)}{|\mathbf{I}_{h_1}| + |\mathbf{I}_{h_2}|}.$$

These results show that, both qualitatively and quantitatively, the difference between measures M_2 and M_3 is not important in many special cases.

Measures M_1, M_2, and M_3 define three comparative notions of truthlikeness (relative to C_t) which, by (T1)–(T7), have intuitively satisfactory properties. Comparisons of truthlikeness among constituents reduce to comparisons of distances given by d (cf. result (T6)), and among true generalizations they reduce, in the case of M_1 and M_2, to comparisons of degrees of information given by I (cf. result (T5)). All measures M_1, M_2, and M_3 make it possible to compare false generalizations for truthlikeness; moreover, a false generalization may even be more truthlike than a true one.[20]

6. ESTIMATION OF TRUTHLIKENESS

A common prejudice against the notion of truthlikeness is the contention that it is useless in the following sense: to know the degree of truthlikeness of a theory S, we have to be able to measure its distance from the truth, but this measurement presupposes that the truth is already known. Who is interested in merely truthlike theories, when we already know the true one? And if we do not know the true theory, then we do not know the truthlikeness of other theories, either.

Popper's quantitative theory of verisimilitude does not tell much about the indicators of truthlikeness: even if his degrees of corroboration depend upon evidence, they have the minimum value -1 for all theories falsified by the evidence.[21] For the same reason, the inductive probabilities of falsified theories cannot serve as estimates of their degrees of truthlikeness.[22] Miller has claimed that "possible-world approaches like that of Hilpinen threaten not to be empiricist enough – there is no indication how experience could influence judgements of truthlikeness at all" (Miller, 1975b, p. 217). It is shown in this section, however, that it is quite easy to supplement the semantical considerations of truthlikeness, in Sections 4 and 5, with reasonable methods of estimating degrees of truthlikeness on the basis of some available evidence. Moreover, these methods meet the objections mentioned above. All we need here is the old horror of the Popperians – a suitable ·system of inductive logic.[23]

To treat the simplest case, assume that a sample of n individuals has been drawn from the domain U such that n_i individuals have been found to exemplify the cell Ct_i $(i = 1, ..., K)$ of language L. Here $n_i \geq 0$ for all $i = 1, ..., K$ and $n_1 + \cdots + n_K = n$. Let c be the number of different cells which are exemplified in the sample, and let e_n^c be a sentence of L which describes this sample. Let C_* be the L-constituent with

$$\mathbf{CT}_* = \{Ct_i \mid n_i > 0\},$$

i.e. C_* is the minimally wide L-constituent compatible with evidence e_n^c. Then a L-constituent C_i is compatible with e_n^c if and only if $\mathbf{CT}_* \subseteq \mathbf{CT}_i$.

Unless all the cells of L are exemplified in e_n^c (i.e. $c = K$), there are more than one L-constituent compatible with e_n^c. On the basis of this evidence, we cannot usually know which L-constituent is the true one. If we knew the true constituent, the above definitions of truthlikeness could directly be applied to the measurement of the truthlikeness of arbitrary generalizations in L. As this possibility is not open to us, the best we can do is to try to evaluate the *expected degree of truthlikeness* of generalizations on the basis of evidence e_n^c. Depending on the choice of measure M_i $(i = 1, 2, 3)$, three different measures ver_i are obtained for the expected truthlikeness:

$$(25) \qquad ver_i(h/e_n^c) = \sum_{j=1}^{2^K} P(C_j/e_n^c)M_i(h, C_j).$$

To obtain the probabilities needed for formula (25), Hintikka's system of inductive logic may be used.[24] With the assumption $\lambda(w) = w$, the probability $P(C_j/e_n^c)$ of a L-constituent C_j of width w_j given evidence e_n^c is defined by the formula

$$(26) \qquad P(C_j/e_n^c) = \frac{\dfrac{(\alpha + w_j - 1)!}{(n + w_j - 1)!}}{\displaystyle\sum_{i=0}^{K-c} \binom{K-c}{i} \dfrac{(\alpha + c + i - 1)!}{(n + c + i - 1)!}}$$

where α is a non-negative real-valued parameter. When α and n are large enough in relation to K, formula (26) can be approximated by

$$(27) \qquad P(C_j/e_n^c) = \frac{\left(\dfrac{\alpha}{n}\right)^{w_j - c}}{\left(1 + \dfrac{\alpha}{n}\right)^{K-c}}.$$

(In formulae (26) and (27), it is assumed that C_j is compatible with e_n^c.) When $n \to \infty$, while α and c are fixed, we have by (27)

$$(28) \qquad P(C_j/e_n^c) \to 1, \quad \text{if} \quad w_j = c$$
$$\to 0, \quad \text{if} \quad w_j > c.$$

In other words, $P(C_j/e_n^c)$ has asymptotically the value one if and only if C_j is the constituent C_*.

It follows from (28) that $ver_i(h/e_n^c)$ is asymptotically equal to the distance $M_i(h, C_*)$ of h from the constituent C_*:

$$(29) \qquad \text{If } n \to \infty \text{ and } c \text{ is fixed, then for all } i = 1, 2, 3$$
$$ver_i(h/e_n^c) \to M_i(h, C_*).$$

In other words, when n is large enough, so that we find it reasonable to assume that e_n^c already exhausts the qualitative variety of the domain U as expressible in language L (i.e. that c will remain constant even if n grows), then the degree of truthlikeness of a generalization h in L may be estimated by guessing that C_* is the true constituent and then evaluating the truthlikeness of h on the basis of this assumption. It follows from (T2) and (29) that

(30) $ver_i(h/e_n^c) \xrightarrow[n \to \infty]{} 1$ iff $\vdash h \equiv C_*$.

In Hintikka's system, C_* is, among L-constituents compatible with e_n^c, the least probable initially and the most probable given e_n^c. It therefore maximizes the difference $P(h/e_n^c) - P(h)$, which thus serves among constituents as an indicator of high truthlikeness. High posterior probability is not sufficient for judging a generalization as highly truthlike – for example, logical truths have the posterior probability one but their estimated degree of truthlikeness is asymptotically only γ or $\frac{1}{2}$.

When n is large enough, the value of $ver_i(C_j/e_n^c)$ for a constituent C_j compatible with e_n^c decreases when the distance $d(C_j, C_*)$ increases. As here

$$d(C_j, C_*) = w_j - c,$$

formula (27) shows that, for sufficiently large n, the posterior probability $P(C_j/e_n^c)$ covaries with the expected degree of truthlikeness of C_j. Thus, for constituents compatible with evidence, posterior probabilities with sufficiently large evidence serve as indicators of expected truthlikeness.

The posterior probability of constituents which are *falsified* by evidence e_n^c is, of course, zero, and therefore cannot serve as an indicator of their truthlikeness. Let c_1 be the number of cells in $\mathbf{CT}_i$ which are exemplified in evidence e_n^c. Then constituent C_i is falsified by e_n^c if and only if $c_1 < c$. The distance between C_i and C_* is then

$$d(C_i, C_*) = (c - c_1) + (w_i - c_1) = c + w_i - 2c_1,$$

so that asymptotically

$$ver_i(C_i/e_n^c) = 1 - \frac{1}{K}(c + w_i - 2c_1).$$

Therefore, the probability $P(C_j/e_n^{c_1})$ for a constituent C_j with the width $w_j = c + w_i - c_1$ serves as an indicator of the truthlikeness of constituent

C_i. Note that if C_i is compatible with e_n^c, then $c_1 = c$ and $P(C_j/e_n^{c_1}) = P(C_i/e_n^c)$.

Let h be a generalization in L which is compatible with evidence e_n^c. Then the degree of *corroboration* of h on evidence e_n^c has been defined by Hintikka (1968) as follows:

$$(31) \qquad \text{corr}\,(h/e_n^c) = \min\,\{P(C_i/e_n^c)\,|\,i \in \mathbf{I}_h(e)\},$$

where $\mathbf{I}_h(e)$ includes the indices of those constituents in the L-normal form of h which are compatible with e_n^c. When n is large enough, measure $\text{corr}\,(h/e_n^c)$ covaries with the measure $I(h, C_*)$. Among generalizations with the same value of $V(h, C_*)$, measure corr serves as an indicator of asymptotic truthlikeness (in the sense of M_1 and M_2). Moreover, if h is a generalization expressed by a universal generalization, then its normal form automatically contains the constituent C_*.[25] Among universal generalizations compatible with the evidence, measure corr is thus a reasonable indicator of truthlikeness.

Estimates of truthlikeness for generalizations falsified by the evidence e_n^c can in principle be calculated from the given formulae.

It has been assumed above that the evidence e_n^c is *complete* with respect to language L in the sense that it is known of each individual in the sample which cell of L it exemplifies. In a more general situation, it may be assumed that the evidence is described in a language L_0 and the generalizations are formulated in a language L which is an extension of L_0. Then there are several constituents C_i^r of L which asymptotically ($n \to \infty$) receive non-zero posterior probabilities on the evidence e_n^c in L_0; all these constituents $C_{i_1}^r, ..., C_{i_t}^r$ of L are compatible with the constituent C_* of L_0 (see Niiniluoto, 1976c). If

$$p_j = \lim_{n \to \infty} P(C_{i_j}^r/e_n^c), \quad \text{for} \quad j = 1, ..., t,$$

where $p_j > 0$ and $p_1 + \cdots + p_t = 1$, then for a generalization h in L

$$(32) \qquad ver_i(h/e_n^c) \xrightarrow[n \to \infty]{} \sum_{j=1}^{t} p_j M_i(h, C_{i_j}^r).$$

In other words, the expected degree of truthlikeness of a generalization h in L given evidence e_n^c in L_0 is asymptotically obtainable by guessing that the generalization

$$(33) \qquad \bigvee_{j=1}^{t} C_{i_j}^r$$

is true and then evaluating the 'distance' of h from (33) as a weighted average of the distances of h from the disjuncts of (33), using the probabilities p_j as weights.

Another generalization of formula (25) is provided by the case where e_n^c and h are expressed in the same language L_0 and, besides e_n^c, there is available theoretical *background knowledge* as expressed by a theory T in an extension L of L_0. The expected degree of truthlikeness of h in L_0 given e_n^c in L_0 and T in L is then defined by

$$(34) \qquad ver_i(h/e_n^c \& T) = \sum_{j=1}^{2^K} P(C_j/e_n^c \& T) M_i(h, C_j).$$

It follows that asymptotically $(n \to \infty)$ $ver_i(h/e_n^c \& T)$ equals the value of $M_i(h, C_j)$, where C_j is the minimally wide L_0-constituent compatible with both e_n^c and T.[26]

7. TRUTHLIKENESS AND SCIENCE

Popper has suggested that "search for verisimilitude is a clearer and a more realistic aim than the search for truth" (Popper, 1972, p. 57). Moreover,

... while we can never have sufficiently good arguments in the empirical sciences for claiming that we have actually reached the truth, we can have strong and reasonably good arguments for claiming that we may have made progress towards the truth. (*Ibid.*, pp. 57–58.)

In contrast with this thesis, we have found in Section 6 that claims of truthlikeness are equally conjectural as claims of truth. And in view of its conceptual relativity and its pragmatic ambiguity truthlikeness can hardly be called a 'clearer' aim than truth.

According to the results of Section 6, expected truthlikeness of generalizations can be increased either by eliminating from normal forms constituents which seem to be far from the true one ('eliminative induction'; cf. Hintikka, 1968) or by adding to normal forms constituents which seem to be close to the true one. In general, however, a scientist who maximizes the expected truthlikeness is eventually lead to the generalization C_*, that is, to the same generalization as the scientist who maximizes expected information content (as measured by cont $(h) = 1 - P(h)$). In this sense, truthlikeness does not essentially differ from informative truth as an epistemic utility.

The notion of truthlikeness is methodologically important mainly for the reason that it shows, in non-instrumentalist terms, why "theories retain their interest even if we have reason to believe that they are false" (Popper, 1972, p. 57). This can be seen especially clearly in a situation where evidence e_n^c (possibly together with background theories) has eliminated all the universal generalizations of the given language L (i.e. $c = K$). Only the L-constituent C_i with $w_i = K$ is then compatible with the evidence, and all the genuine theories expressible in L are believed to be false. Some of these false theories may nevertheless be more 'interesting' than others (or the true L-constituent C_i itself!) in the sense that they are closer to the truth at some level of analysis deeper than L.[27]

The notion of truthlikeness which has been studied in this paper is a variant, or special case, of a more general notion of truthlikeness which takes into account K different respects of comparison each of which admits several degrees. According to Hilpinen's (1976) definition, 'approximate truth' means 'close to the truth in all respects of comparison', while the straightforward generalization of the definition of this paper corresponds to the idea 'true in most respects of comparison'. This latter idea seems to have a great methodological significance, since many *idealizations* in science have precisely the characteristic that in most respects they correspond to the truth (at least very closely) but some of the relevant variables are yet assumed to have an extreme value which is very far from the truth.[28]

As the concluding remark, note that with simple modifications of the above definitions we obtain a number of other notions of truthlikeness which have a wide range of methodological applications. For example, if x_i and y_i indicate the number of individuals in cell Ct_i, then formula (7) defines the distance between two 'structural descriptions' of the given domain. The distance between two Ct-predicates of L can be defined in analogy with formula (7). As each complex predicate expressible in L is equivalent to a finite disjunction of Ct-predicates, a measure of truthlikeness for the *singular sentences* in L is obtained in the same way as measures M_i are defined above. This definition – as well as the definition of truthlikeness for generalizations – can easily be generalized to situations which involve several families of attributes with a similarity metric on each family (in the sense of Carnap).

University of Helsinki

NOTES

[1] See Laudan (1966), p. 94.
[2] For details and references, see Laudan (1973).
[3] Cf. the remarks in Niiniluoto (1976a).
[4] For a suggestive defense of an alternative view, see Kuhn (1970), pp. 205–207. According to Kuhn, scientific theories are "better instruments for discovering and solving puzzles" than their predecessors, but not somehow better representations of "what nature is really like".
[5] It need not be claimed that the approach to the truth is uniform, that is, displayed by each succession of two theories. However, the realists at least claim that some successions of theories, as well as the overall development of science, show genuine 'progress' in an ontological sense.
[6] For Popper's theory of verisimilitude, see Popper (1963), pp. 231–237, 391–403; Popper (1972), 47–60, 81, 103, 334–335; and Popper (1974), pp. 1100–1103. Popper's reference to Quine's remarks on Peirce is made in Popper (1974), p. 1101.
[7] See Miller (1974), Tichý (1974), and Harris (1974). Hattiangadi (1975) informs that C. G. Hempel had proved essentially the same negative result already in 1970; the given proof is valid for finitely axiomatizable theories.
[8] See Harris (1976).
[9] The definition of truthlikeness involves a measure of the 'distance' between two possible worlds. This measure is specified here only in the special case where the two possible worlds are described by means of generalizations in a monadic first-order language. The results which essentially depend upon this measure are not applicable to situations which involve relational concepts or quantitative concepts – therefore the notion of truthlikeness in this paper does not directly correspond to the notion of approximate truth. (See Section 9, however.) It should be noted, however, that many results of this paper are independent of the choice of the distance measure.
[10] The Ct-predicates Ct_j, $j = 1, ..., K$, of language L are defined by conjunctions of the form

$$(\pm)O_1(x) \, \& \, \cdots \, \& \, (\pm)O_k(x),$$

where the symbol '($\pm$)' may be replaced by the negation sign $\sim$ or nothing. When language L is interpreted on a domain U, the Ct-predicates Ct_j partition U into K Cells – each member of U belongs to one and only one cell of this partition. A L-constituent C_i is a statement which specifies that certain cells are instantiated in U – let their set be $\mathbf{CT}_i$ – and the other cells are empty. Thus,

$$C_i = \bigvee_{Ct_j \in \mathbf{CT}_i} (Ex)Ct_j(x) \ \& \ (x)\left[\bigwedge_{Ct_j \in \mathbf{CT}_i} Ct_j(x) \right].$$

The number $w_i = \text{card}\,(\mathbf{CT}_i)$ is called the *width* of constituent C_i. The number of different L-constituents is 2^K. Constituents are the strongest generalizations (i.e. quantificational statements without individual constants) expressible in language L. They are logically incompatible with each other and logically exhaustive: one and only one L-constituent C_t is true. For these notions, see for example Niiniluoto and Tuomela (1973) and Niiniluoto (1976b).

[11] The situation is similar, but more complex, if full first-order languages with relation symbols are considered. Corresponding to the sequence (4), an infinite sequence L_i, $i = 1, 2, ...$, of full first-order languages with vocabularies $\lambda_1 \subseteq \lambda_2 \subseteq \cdots$ defines the following diagram:

$$
\begin{array}{cccc}
B_1^{(0)} \leqslant B_2^{(0)} \leqslant \ldots \leqslant B_i^{(0)} \leqslant \ldots \\
\text{\tiny A} \quad \text{\tiny A} \qquad \text{\tiny A} \\
B_1^{(1)} \leqslant B_2^{(1)} \leqslant \ldots \leqslant B_i^{(1)} \leqslant \ldots \\
\text{\tiny A} \quad \text{\tiny A} \qquad \text{\tiny A} \\
\vdots \quad \vdots \qquad \vdots \\
\text{\tiny A} \quad \text{\tiny A} \qquad \text{\tiny A} \\
B_1^{(d)} \leqslant B_2^{(d)} \leqslant \ldots \leqslant B_i^{(d)} \leqslant \ldots \\
\text{\tiny A} \quad \text{\tiny A} \qquad \text{\tiny A} \\
\vdots \quad \vdots \qquad \vdots
\end{array}
$$

Here $B_i^{(d)}$ is the set of constituents of language L_i with depth d; the quantificational depth d indicates the number of individuals which are considered in relation to each other in constituents at this depth. See Hintikka (1973) for the general theory of first-order constituents.

[12] Note that according to the measures M_1, M_2, and M_3 of truthlikeness (to be defined in Section 5 below) we have

$$M_i(C_1^1, C_t^1) = 0 < 1 = M_i(C_2^1, C_t^1) \qquad (i = 1, 2, 3)$$

but

$$M_i(C_1^1, C_t^2) = \tfrac{1}{2} = M_i(C_2^1, C_t^2) \qquad (i = 2, 3)$$

$$M_3(C_1^1, C_t^2) = \tfrac{1}{2} \quad \text{and} \quad M_3(C_2^1, C_t^2) = \gamma$$

and

$$M_i(C_1^1, C_t^3) = \tfrac{3}{4} > \tfrac{1}{2} = M_2(C_2^1, C_t^3) \qquad (i = 1, 2, 3)$$

$$> \tfrac{29}{60} = M_3(C_2^1, C_t^3)$$

$$M_3(C_2^1, C_t^3) = \gamma.$$

Thus, according to measures M_2 and M_3, constituent C_2^1 is more truthlike than constituent C_1^1 at the level L_1, equally truthlike at the level L_2, and less truthlike at the level L_3.

[13] For discussions about 'conservative' and 'radical' conceptual shifts, see Chapter 10 of Niiniluoto and Tuomela (1973) and Niiniluoto (1976b).

[14] See Stalnaker (1968).

[15] Hilpinen (1976) follows recent theories of conditionals (see Stalnaker (1968) and Lewis (1973)) by assuming the overall similarity between possible worlds as a primitive comparative notion. In this paper, overall comparisons of possible worlds are reduced to comparisons of constituents (i.e. sentences which tell about the world everything which is relevant as far as generalizations are concerned). In virtue of this reduction, it is possible to consider in detail the problem of specifying the similarity metric. Constituents of full first-order languages are not considered in this paper – all attempts to measure their 'distance' seem to involve assumptions about the relative importance of the various quantificational depths.

[16] Cf. Hilpinen's (1976) results for his set-function E.

[17] Cf. Hilpinen's (1976) results for his set-function I.

[18] This observation is made in Hilpinen (1976). He proposes only a comparative notion of truthlikeness – it corresponds to the following definition: h_1 is *more truthlike* than h_2 if and only if $V(h_1, C_t) \geq V(h_2, C_t)$ and $I(h_1, C_t) \geq I(h_2, C_t)$, where at least one of these inequalities is strict.

[19] A proposal to this effect has been made by Tichý (1974), p. 159. He specifies a distance measure between constituents of propositional logic only; this measure is not very plausible, however (see Miller, 1975b, p. 214).

[20] It is desirable, for example, that a false but informative theory may be more truthlike than a logical truth. According to the definition given in Note 18, a true theory is either more truthlike than a given false theory or incomparable with it.

[21] For Popper's suggestion that degrees of corroboration may serve as indicators of verisimilitude, see Popper (1972), p. 103.

[22] The inductive probability $P(h/e)$ indicates a rational degree of belief in the *truth* of h given evidence e. It has been argued by L. J. Cohen that since theories may be supported by evidence which falsifies them, support cannot be measured by inductive probabilities (see Cohen, 1973). Support in this sense can be measured by estimated degrees of truthlikeness.

[23] The possibility of developing a theory of the estimation of truthlikeness on the basis of inductive logic strikes to me as a convincing proof that inductive logic is not a "degenerating research programme" (cf. Lakatos, 1974, p. 259) or "an unrewarding feat of formalization" (cf. Miller, 1972, p. 51, n2).

[24] See Hintikka (1966) or Niiniluoto and Tuomela (1973).

[25] A universal generalization claims, in effect, that certain cells of L are empty, so that its normal form contains *all* the constituents of L which leave these cells empty. Hintikka's measure of corroboration is applied to universal generalizations in Niiniluoto and Tuomela (1973), pp. 128–138.

[26] Probabilities of the form $P(C_j/e_n^c \& T)$ are studied in Chapters 3–5 of Niiniluoto and Tuomela (1973).

[27] Cf. Figure 2 in Section 3. To give a specific example, let L_1 be a language with $k = 2$, and interpret the predicates O_1 and O_2 as designating the properties of being a swan and being white, respectively. Suppose that white swans, non-white swans, white non-swans,

and non-white non-swans have been observed, so that the L_1-constituent which claims that all the four cells of L_1 are exemplified is believed to be true. At the level of L_1, this constituent C_1 is more truthlike than the false constituent C_2 which claims that all swans are white. Let L_2 be the language with the predicates O_1, O_2, and O_3, where O_3 is interpreted as designating the property of being black. In view of the 'meaning postulate' that no object is both white and black, there are six cells of language L_2. As all swans are either white or black, the true L_2-constituent C_t^2 claims that all the cells of L_2 but O_1 & $\sim O_2$ & $\sim O_3$ are instantiated. Then C_1 is equivalent to a disjunction of nine L_2-constituents, while C_2 is equivalent to a disjunction of three L_2-constituents. At the level of L_2, measures M_i, $i = 1$, 2, 3, of truthlikeness give the following values to C_1:

$$M_1(C_1, C_t^2) = \gamma + \tfrac{3}{6}(1-\gamma) = \tfrac{1}{2}(1+\gamma)$$
$$M_2(C_1, C_t^2) = \tfrac{3}{4}$$
$$M_3(C_1, C_t^2) = 1 - \tfrac{1}{6} \cdot \tfrac{15}{9} = \tfrac{13}{18}$$

and the following values to C_2:

$$M_1(C_2, C_t^2) = \tfrac{5}{6}\gamma + \tfrac{4}{6}(1-\gamma) = \tfrac{1}{6}(4+\gamma)$$
$$M_2(C_2, C_t^2) = \tfrac{3}{4}$$
$$M_3(C_2, C_t^2) = 1 - \tfrac{1}{6} \cdot \tfrac{5}{3} = \tfrac{13}{18}.$$

Thus, according to measures M_2 and M_3, C_1 and C_2 are equally truthlike at the level of L_2. Moreover, if the information-factor has more weight than the truth-factor (i.e. $\gamma < \tfrac{1}{2}$), then measure M_1 gives to C_2 a higher degree of truthlikeness than to C_1 at the level of L_2.

[28] Cf. Leszek Nowak's (1975) interesting account of scientific progress through the 'concretization' of idealizational laws, that is, through the elimination of idealizational assumptions.

BIBLIOGRAPHY

Cohen, L. J.: 1973, 'The Paradox of Anomaly', in R. J. Bogdan and I. Niiniluoto (eds.), *Logic, Language, and Probability*, D. Reidel, Dordrecht and Boston, 1973, pp. 78–82.
Harris, J. H.: 1974, 'Popper's Definition of "Verisimilitude" ', *The British Journal for the Philosophy of Science* **25**, 160–166.
Harris, J. H.: 1976, 'On Comparing Theories', *Synthese* **32**, 29–76.
Hattiangadi, J. N., 'After Verisimilitude', in *Contributed Papers, 5th International Congress of Logic, Methodology and Philosophy of Science*, London, Ontario, 1975, pp. V–49–50.
Hilpinen, R.: 1976, 'Approximate Truth and Truthlikeness', in M. Przelecki and R. Wojcicki (eds.), *Proceedings of the Conference of Formal Methods in the Methodology of the Empirical Sciences, Warsaw, 1974*, D. Reidel, Dordrecht and Boston.

Hintikka, K. J.: 1966, 'A Two-Dimensional Continuum of Inductive Methods', in K. J. Hintikka and P. Suppes (eds.), *Aspects of Inductive Logic*, North-Holland, Amsterdam, pp. 113–132.

Hintikka, K. J.: 1968, 'Induction by Enumeration and Induction by Elimination', in I. Lakatos (ed.), *The Problem of Inductive Logic*, North-Holland, Amsterdam, pp. 191–216.

Hintikka, K. J.: 1973, *Logic, Language-Games, and Information: Kantian Themes in the Philosophy of Logic*, Oxford University Press, Oxford.

Kuhn, T.: 1970, *The Structure of Scientific Revolutions*, 2nd ed., University of Chicago Press, Chicago.

Lakatos, I.: 1974, 'Popper on Demarcation and Induction', in P. A. Schilpp (ed.), *The Philosophy of Karl Popper* (The Library of Living Philosophers, Vol. XIV), Open Court, La Salle, Ill., pp. 241–273.

Laudan, L.: 1966, 'The Clock Metaphor and Probabilism: The Impact of Descartes on English Methodological Thought, 1650–65', *Annals of Science* **22**, 73–104.

Laudan, L.: 1973, 'Peirce and the Trivialization of the Self-Correcting Thesis', in R. N. Giere and R. S. Westfall (eds.), *Foundations of Scientific Method: The Nineteenth Century*, Indiana University Press, Bloomington, pp. 275–306.

Lewis, D.: 1973, *Counterfactuals*, Blackwell, Oxford.

Miller, D.: 1972, 'The Truth-Likeness of Truthlikeness', *Analysis* **33**, 50–55.

Miller, D.: 1974, 'Popper's Qualitative Theory of Verisimilitude', *The British Journal for the Philosophy of Science* **25**, 166–177.

Miller, D.: 1975a, 'The Accuracy of Predictions', *Synthese* **30**, 159–191.

Miller, D.: 1975b, 'The Accuracy of Predictions: A Reply', *Synthese* **30**, 207–219.

Niiniluoto, I.: 1976a, 'Notes on Popper as Follower of Whewell and Peirce', forthcoming in *Ajatus* **37**.

Niiniluoto, I.: 1976b, 'Inquiries, Problems, and Questions: Remarks on Local Induction', in R. J. Bogdan (ed.), *Local Induction*, D. Reidel, Dordrecht and Boston, pp. 263–296.

Niiniluoto, I.: 1976c, 'Theoretical Concepts and Inductive Logic', in M. Przelecki and R. Wojcicki (eds.), *Proceedings of the Conference of Formal Methods in the Methodology of the Empirical Sciences, Warsaw, 1974*, D. Reidel, Dordrecht and Boston.

Niiniluoto, I. and Tuomela, R.: 1973, *Theoretical Concepts and Hypothetico-Inductive Inference*, D. Reidel, Dordrecht and Boston.

Nowak, L.: 1975, 'Relative Truth, The Correspondence Principle and Absolute Truth', *Philosophy of Science* **42**, 187–202.

Popper, K. R.: 1963, *Conjectures and Refutations: The Growth of Scientific Knowledge*, Routledge and Kegan Paul, London.

Popper, K. R.: 1972, *Objective Knowledge: An Evolutionary Approach*, Oxford University Press, Oxford.

Popper, K. R.: 'Replies to My Critics', in P. A. Schilpp (ed.), *The Philosophy of Karl Popper* (The Library of Living Philosophers, Vol. XIV), Open Court, La Salle, Ill., pp. 961–1197.

Stalnaker, R. C.: 1968, 'A Theory of Conditionals', in N. Rescher (ed.), *Studies in Logical Theory* (APQ Monograph No. 2), Blackwell, Oxford.

Tichý, P.: 1974, 'On Popper's Definition of Verisimilitude', *The British Journal for the Philosophy of Science* **25**, 155–160.

WESLEY C. SALMON*

A THIRD DOGMA OF EMPIRICISM

In 1951, W. V. Quine published his provocative and justly famous article, 'Two Dogmas of Empiricism'. At about the time Quine was mounting this attack, a number of 'empiricists' were busily establishing what has subsequently become, in my opinion, a third dogma. The thesis can be stated quite succinctly: *scientific explanations are arguments*. This view was elaborated at considerable length by a variety of prominent philosophers, including R. B. Braithwaite (1953), Ernest Nagel (1961), Karl Popper (1959) and most especially Carl G. Hempel.[1] Until the early 1960s, although passing mention was sometimes made of the need for inductive explanation, attention was confined almost exclusively to deductive explanation. In 1962, however, Hempel (1962a) made the first serious attempt to provide a detailed analysis of inductive (or statistical) explanation. In that same year, in a statement referring explicitly to both deductive and inductive explanations, he characterized the 'explanatory account' of a particular event as 'an *argument* to the effect that the event to be explained ⋯ was to be expected by reason of certain explanatory facts' (Hempel, 1962b, my italics). Shortly thereafter he published an improved and more detailed version of his treatment of inductive-statistical (I-S) explanation (Hempel, 1965, pp. 381–412). In this newer discussion, as well as in many other places, Hempel has often reiterated the thesis that explanations, both deductive and inductive, are arguments.[2] The purpose of this present paper is to raise doubts about the tenability of that general thesis by posing three questions – ones which will, I hope, prove embarrassing to those who hold it.[3]

QUESTION 1. Why are irrelevancies harmless to arguments but fatal to explanations?

In deductive logic, irrelevant premises are pointless, but they do not undermine the validity of the argument. Even in the relevance logic of

Anderson and Belnap, $p \& q \vdash p$ is a valid schema. If one were to offer
the argument,

> All men are mortal.
> Socrates is a man.
> Xantippe is a woman.
> Socrates is mortal.

it would seem strange, and perhaps mildly amusing, but its logical
status would not be impaired by the presence of the third premise.
There are more serious examples. When it was discovered that the
axioms of the propositional calculus in *Principia Mathematica* were not
all mutually independent, there was no thought that the logical system
was thereby vitiated. Nor is the validity of Propositions 1–26 of Book I
of Euclid called into question as a result of the fact that they all follow
from the first four postulates alone, without invoking the famous fifth
(parallel) postulate. This fact, which has important bearing upon the
relationship between Euclidean and non-Euclidean geometries, does
not represent a fault in Euclid's deductive system.

When we turn to deductive explanations, however, the situation is
radically different. The rooster who explains the rising of the sun on
the basis of his regular crowing is guilty of more than a minor logical
inelegancy. So also is the person who explains the dissolving of a piece
of sugar by citing the fact that the liquid in which it dissolved is *holy*
water. So also is the man who explains *his* failure to become pregnant
by noting that he has faithfully consumed birth control pills.[4]

The same lack of parity exists between inductive arguments and
explanations. In inductive logic there is a well-known requirement of
total evidence.[5] This requirement demands the inclusion of all *relevant*
evidence. Since irrelevant 'evidence' has, by definition, no effect upon
the probability of the hypothesis, inclusion of irrelevant premises in an
inductive argument can have no bearing upon the degree of strength
with which the conclusion is supported by the premises.[6] If facts of
unknown relevance turn up, inductive sagacity demands that they be
mentioned in the premises, for no harm can come from including them
if they are irrelevant, but considerable mischief can accrue if they are
relevant and not taken into account.

When we turn our attention from inductive arguments to inductive explanations, the situation changes drastically. If the consumption of massive doses of vitamin C is irrelevant (statistically) to immunity to the common cold, then 'explaining' freedom from colds on the basis of use of that medication is worse then useless.[7] So also would be the 'explanation' of psychological improvement on the basis of psychotherapy if the spontaneous remission rate for neurotic symptoms were equal to the percentage of 'cures' experienced by those who undergo the particular type of treatment.[8]

Hempel recognized from the beginning the need for some sort of requirement of total evidence for inductive-statistical explanation; it took the form of the *requirement of maximal specificity* (Hempel, 1965, pp. 394–403). This requirement stipulates that the reference class to which an individual is referred in a statistical explanation be narrow enough to preclude, in the given knowledge situation, further relevant subdivision. It does not, however, prohibit irrelevant restriction. I have therefore suggested that this requirement be amended as the *requirement of the maximal class of maximal specificity* (Salmon, 1970, § 5) This requirement demands that the reference class be determined by taking account of all relevant considerations, but that it not be irrelevantly partitioned.

Inference, whether inductive or deductive, demands a requirement of total evidence – a requirement that *all* relevant evidence be mentioned in the premises. This requirement, which has substantive importance for inductive inferences, is automatically satisfied for deductive inferences. Explanation, in contrast, seems to demand a further requirement – namely, that *only* considerations relevant to the explanandum be contained in the explanans. This, it seems to me, constitutes a deep difference between explanations and arguments.

Question 2 comes in two distinct forms, which I shall number 2 and 2′ respectively. The two forms may actually express different questions, but they are so closely related as to deserve some sort of intimate linkage.

QUESTION 2.   Can events whose probabilities are low be explained?

Although they made no attempt to provide an explication of inductive-statistical explanation in their classic 1948 paper, Hempel and Oppenheim did acknowledge the need for explanations of that sort

(Hempel, 1965, pp. 250–251). On Hempel's subsequent account of inductive-statistical explanation, events whose probabilities are high (relative to a suitably specified body of knowledge) are amenable to explanation. A high probability is demanded by the requirement that the explanation be an argument to the effect that the event in question *was to be expected,* if not with certainty, then with high probability, in virtue of the explanatory facts.

If some events are probable, without being certain, others are improbable. If a coin has a strong bias for heads, say 0.9, then tails has a non-vanishing probability, and a small percentage of the tosses will in fact result in tails. It seems strange to say that the results of tosses in which the coin lands heads up can be explained, while the results of those tosses of the very same coin in which tails show are inexplicable. To be sure, the head-outcomes far outnumber the tail-outcomes, but is it not an eccentric prejudice which leads us to discriminate against the minority, condemning its members to the realm of the inexplicable?

The case need not rest on examples of the foregoing sort. In a number of well-known examples we seem to be able to offer genuine explanations of events whose non-occurrence is more probable than not. Michael Scriven has pointed out that the probability of paresis developing in cases of latent untreated syphilis is quite small, but syphilis is accepted as the explanation of paresis in those cases in which it does occur.[9] Similarly, as I understand it, mushroom poisoning may afflict only a small percentage of individuals who eat a particular type of mushroom, but the eating of the mushroom would unhesitatingly be offered as the explanation in instances of the illness in question.[10] Moreover, a uranium nucleus may have a probability as low as 10^{-38} of decaying by spontaneously ejecting an alpha-particle at a particular moment. When decay does occur we explain it in terms of the 'tunnel effect', which assigns a low probability to that event.

Imposition of the 'high probability requirement' upon explanations produces a serious malady that Henry Kyburg (1970) has dubbed 'conjunctivitis'.[11] Because of the basic multiplicative rule of the probability calculus, the joint occurrence of two events is normally less probable than either event individually. This is illustrated by the above-mentioned biased coin. The probability of heads on any given toss is 0.9, while the probability of two heads in a row is 0.81. If 0.9

were the minimal value acceptable in inductive-statistical explanation, we would be able to explain each of the two tosses separately, but their joint occurrence would be unexplainable. The moral to be drawn from examples of this kind is, it seems to me, that there is no reasonable way of answering the question, "How high is high enough?"

If conjunctions are the enemies of high probabilities, disjunctions are their indispensable allies. If a fair coin is tossed 10 times, there is a probability of 1/1024 that it will come up heads on all ten tosses. This sequence of events constitutes a (complex) low probability event, and as such, it is unexplainable. Even if the outcome is 5 heads and 5 tails, however, the probability of that sequence of results, *in the particular order in which they occurred*, is also 1/1024. It too, is a low probability event, and as such, is unexplainable. If, however, we consider the probability of 5 heads and 5 tails *regardless of order*, we are considering the disjunction of all of the 252 distinct orders in which that outcome can occur. Even this extensive disjunction has a probability of only about 0.246; hence, even it fails to qualify as a high probability event. If, however, we consider the probability of getting almost one-half heads in 10 tosses, i.e. 4 or 5 or 6 heads, this disjunction is ample enough to have a probability somewhat greater than 0.5 (appx. 0.656). The general conclusion would seem to be that, for even moderately complex events, every specific outcome has a low probability, and is consequently incapable of being explained.[12] The only way to achieve high probabilities is to erase the specific character of the complex event by disjunctive dilution.

Richard Jeffrey (1969) and James Greeno (1970) have both argued, quite correctly, I believe, that the degree of probability assigned to an occurrence in virtue of the explanatory facts is not the primary index of the value of the explanation. Suppose, for example, that two individuals, Sally Smith and John Jones, both commit suicide. Using our best psychological theories, and summoning all available relevant information about both persons (such as sex, age, race, state of health, marital status, etc.), we find that there is a low probability that Sally Smith would commit suicide, whereas there is a high probability that John Jones would do so. This does not mean that the explanation of Jones's suicide is better than that of Smith's, for exactly the same theories and relevant factors have to be taken into account in both.

According to an alternative account of statistical explanation, the statistical-relevance (S-R) model, elaborated in Salmon *et al.* (1971), an explanation consists not in an argument but in an assemblage of relevant considerations. On this model, high probability is not the desideratum; rather, the amount of relevant information is what counts. According to the S-R model, a statistical explanation consists of a probability distribution over a homogeneous partition of an initial reference class. A homogeneous partition is one which does not admit of further relevant subdivision.

The subclasses in the partition must also be maximal – that is, the partition must not involve any irrelevant subdivisions. The goodness, or epistemic value, of such an explanation is measured by the gain in information provided by the probability distribution over the partition.[13] If one and the same probability distribution over a given partition of a reference class provides the explanations of two separate events, one with a high probability and one with a low probability, the two explanations are equally valuable.

This approach to statistical explanation offers a pleasant dividend. If we insist that the explanation incorporate the probability distribution over the entire partition – not just the probability value associated with the particular cell of the partition into which the event to be explained happens to fall – we are invoking the statistical analogues of both sufficient and necessary conditions, rather than sufficient conditions alone.[14] This feature of the statistical-relevance model overcomes one severe difficulty experienced by Hempel's deductive-nomological and inductive-statistical models in connection with functional explanations, for these models seem always to demand a sufficient condition where the functional explanation itself provides a necessary condition. Although the task has not yet been accomplished, the statistical-relevance model gives promise of providing an adequate account of functional explanations – a type of explanation that has constituted an embarrassment to the standard inferential approach to explanation (Hempel, 1959).

There is a strong temptation to respond to examples such as the biased coin, paresis, and mushroom poisoning (as well as functional explanations in general) by relegating them to the status of explanation sketches or incomplete explanations. We are apt to believe – often on good grounds – that further investigation would provide the means to say why it is that this syphilitic develops paresis while that one does not, or why one person has an allergic response to a particular type of mushroom while the vast majority of people do not.[15] Such a tack runs the risk, however, of seducing us into the supposition that *all* inductive-statistical explanations are incomplete. It seems to me that we must ask, however, what to say if *not all* examples of low-probability events are amenable to that approach. This problem leads to another question, so closely related to our second question as to be hardly more than a reformulation of it:

QUESTION 2′. Is genuine scientific explanation possible if indeterminism is true?

The term 'determinism' is unquestionably ambiguous. On one plausible construal it can be taken to mean that there are no genuinely homogeneous references classes except in the limiting cases when all A are B or no A are B. Let A be a certain reference class within which the attribute B is present in some but not all cases. According to this version of determinism there must be a characteristic C in terms of which the class A can be partitioned so that within the subclass $A \cap C$ every element is B and within $A \cap \bar{C}$ every element is $\bar{B}$. Suppose, for example, that an alpha-particle approaches a potential barrier with a certain non-vanishing probability of tunneling through and a certain non-vanishing probability of being reflected back. This form of determinism asserts that there is a characteristic present in some cases and absent in others that 'determines' whether the alpha-particle tunnels through or not. This, I take it, is the thesis of hidden variable theorists.

In order to protect this version of determinism from complete trivialization, it is necessary to place some restrictions upon the sort of characteristic C to which we may appeal for purposes of partitioning A. In particular, we must not allow C to be identified with B itself, or any other property whose presence or absence cannot even in principle be ascertained without discovering whether B is present or absent. If, for example, we are discussing the probability of drawing a red ball from an urn, we may partition the class of draws in terms of draws made by males vs draws by females, or draws made in the daytime vs draws made at night; we may not partition the class of draws in terms of draws resulting in a red ball vs draws resulting in other colors, or draws of balls with a color at the opposite end of the visible spectrum from violet vs draws resulting in colors located in other regions of the spectrum.

The problem we confront in attempting to put appropriate restrictions upon the attributes permitted in partitioning reference classes is familiar from another context. In order to implement his definition of a 'collective', Richard von Mises (1964, Chap. I) introduced the notion of a place selection. Although his original explication of the concept of a place selection was certainly unsatisfactory, he did, I believe, correctly identify the explanandum. Subsequent work has made it possible to supply a serviceable definition of 'place selection', and to apply it to the definition of 'homogeneous reference class'.[16] It then remains an open factual question whether there are non-trivial cases of homogeneous reference classes.[17]

In his published writings Hempel is, I believe, committed to the opposite view, for he categorically asserts that inductive-statistical explanations are *essentially* relativized to knowledge situations; he calls this thesis the "epistemic relativity of inductive-statistical explanation" (Hempel, 1965, p. 402). He maintains that, although a reference class which satisfies the requirement of maximal specificity is one which we do not know how to partition relevantly, it is in principle capable of further relevant subdivision in the light of additional knowledge.[18] If there were an inductive-statistical explanation whose lawlike statistical premise involved a genuinely homogeneous reference class – one which, even in principle, could not be further relevantly subdivided – then we would have an instance of an inductive-statistical explanation *simpliciter*, not merely an inductive-statistical explanation *relative to a specific knowledge situation.* Since there are no inductive-statistical explanations *simpliciter* on Hempel's view, he must deny the existence of genuinely homogeneous reference classes, except in trivial cases. In the trivial cases we do not have to rest content with inductive-statistical explanations, for universal laws are available by means of which to construct deductive-nomological explanations. In the ideal limit of complete knowledge, inductive-statistical explanation would have no place, for every explanation would be deductive-nomological.[19]

The relationship between inductive-statistical explanations and deductive-nomological explanations closely parallels the relationship between enthymemes and valid deductive arguments. Since an enthymeme is, by definition, an argument with missing premises, there can be no such thing as a valid enthymeme. Enthymemes can be made to approach validity, we might say, by supplying more and more of the missing premises, but the moment a set of premises sufficient for validity is furnished the argument ceases to be an enthymeme and automatically becomes a valid deductive argument.

Much the same sort of thing can be said about inductive-statistical explanations. The reference class that occurs in a given inductive-statistical explanation and fulfills the requirement of maximal specificity is not genuinely homogeneous; it is still possible in principle to effect a relevant partition, but in our particular knowledge situation we do not happen to know how. As we accumulate further knowledge, we may be able to make further relevant partitions of our reference class, but as long as we fall short of universal laws we have not exhausted all possible relevant information. Progress in constructing inductive-statistical explanations would thus seem to involve a process of closer and closer approximation to the deductive-nomological ideal. Failure to achieve this ideal would not be a result of the non-existence of relevant factors sufficient to provide universal laws; failure to achieve deductive-nomological explanations can only result from our ignorance.

As a result of the foregoing considerations, as well as other arguments advanced by J. A. Coffa (1974), I am inclined to conclude that Hempel's concept of *epistemic relativity of statistical explanations,* which demands relativization of *every* such explanation to a knowledge situation (Hempel, 1965, p. 402), means that Hempel's account of inductive-statistical explanation is completely parasitic upon the concept of deductive-nomological explanation. If, however, indeterminism is true, on any reasonable construal of that doctrine with which I am acquainted, then some reference classes will be actually, objectively, genuinely homogeneous in cases where no universal generalization is possible. In that case, it seems to me, we must have a full-blooded account of inductive-statistical explanation – or statistical explanation, at any rate – which embodies homogeneity of reference classes *not relativized* to any knowledge situation. I do not know whether indeterminism is true; I think we have good physical reasons for supposing it may be true. But regardless of whether indeterminism is true, we need an explication of scientific explanation which is neutral regarding that issue. Otherwise, we face the dilemma of either (1) ruling indeterminism out *a priori* or (2) holding that events are explainable only to the extent that they are fully determined. Neither alternative seems acceptable: (1) the truth or falsity of indeterminism is a matter of physical fact, not to be settled *a priori*, and (2) even if the correct interpretation of quantum mechanics is indeterministic, it still must be admitted to provide genuine scientific explanations of a wide variety of phenomena.

In dealing with Question 2′, I have said quite a bit about determinism and indeterminism without mentioning causal relations. This omission must be corrected. Consideration of the third question will rectify the situation.

QUESTION 3:   Why should requirements of temporal asymmetry be imposed upon explanations (while arguments are not subject to the same constraints)?

A particular lunar eclipse can be predicted accurately, using the laws of motion and a suitable set of initial conditions holding prior to the eclipse; the same eclipse can equally well be retrodicted using posterior conditions and the same laws. It is intuitively clear that, if explanations are arguments, then only the predictive argument can qualify as an explanation, and not the retrodictive one. The reason is obvious. We explain events on the basis of antecedent causes, not on the basis of subsequent effects (or other subsequent conditions).[20] A similar moral can be drawn from Sylvan Bromberger's flagpole example. Given the elevation of the sun in the sky, we can infer the length of the shadow from the height of the flagpole, but we can just as well infer the height of the flagpole from the length of the shadow. The presence of the flagpole explains the occurrence of the shadow; the occurrence of the shadow does not explain the presence of the flagpole. At first blush we might be inclined to say that this is a case of coexistence; the flagpole and the shadow exist simultaneously. Upon closer examination, however, we realize that a causal process is involved, and that the light from the sun must either pass or be blocked by the flagpole *before* it reaches the ground where the shadow is cast.

There are, of course, instances in which inference enjoys a preferred temporal direction. As I write, the July Fourth weekend approaches. We can predict with confidence that many people will be killed, and perhaps give a good estimate of the number. We cannot, with any degree of reliability, predict the exact number – much less the identity of each of the victims. By examining next week's newspapers, however, we can obtain an exact account of the number and identities of these victims, as well as a great deal of information about the circumstances of their deaths.[21] By techniques of dendrochronology (tree ring dating), for another example, relative annual rainfall in parts of Arizona is known for some 8000 years into the past. No one could hazard a reasonable guess about relative annual rainfall for even a decade into the future.

Such examples show that the temporal asymmetry reflected by inferences is precisely the opposite to that exhibited in explanation. We have many records, natural and man-made, of events that have happened in the past; from these records we can make reliable inferences into the past. We do not have similar records of the future.[22] Prognostication is far more difficult than retrodiction; it has no aid comparable to records. No one would be tempted to 'explain' the accidents of a holiday weekend on the basis of their being reported in the newspaper. No one would be tempted to 'explain' the rainfall of past millenia on the basis of the rings in trees of bristlecone pine. If it is indeed true that being an argument is an essential characteristic of scientific explanations, how are we to account for the total disparity of temporal asymmetry in explanations and in arguments? This is a fundamental question for supporters of the inferential view of explanation.

If one rejects the inferential view of scientific explanation, it seems to me that straightforward answers can be given to the foregoing three questions. On the statistical-relevance model of explanation, an explanation is an assemblage of factors that are statistically relevant to the occurrence of the explanandum-event. To offer an item as relevant when it is, in fact, irrelevant is clearly inadmissible. We thus have an immediate answer to our first question, "Why are irrelevancies harmless to arguments but fatal to explanations?"

The second question, in its first form, receives an equally simple and direct answer. Since additional relevant information may raise or lower probabilities, and since assemblages of relevant information may yield high, middling, or low probabilities for an event of a particular sort, the statistical-relevance model has no problems with low probabilities. It never has to face the question, "How high is high enough?" It is absolutely immune to conjunctivitis.

Question 2′ poses the problem, mentioned above, of characterizing homogeneity in an objective and unrelativized manner. Resolution of this problem must be reserved for another occasion, but there seems no reason to doubt that it can be done.[23]

When we come to the third question, regarding temporal asymmetry, we cannot avoid raising the issue of causation. In the classic 1948 article, Hempel and Oppenheim suggested that deductive-nomological explanations are causal explanations, but in subsequent years Hempel has backed away from this position, explicitly dissociating 'covering law' from causal explanations.[24] The time has come, it seems to me, to put the 'cause' back into 'because'. Consideration of the temporal asymmetry issue forces reconsideration of causation in explanation.

There are two levels of explanation in the statistical-relevance model. At the first level, we invoke statistical regularities to provide a relevant partition of a given reference class into maximal homogeneous subclasses. For example, we place Sally Smith in a subclass of Americans which is defined in terms of characteristics such as age, sex, race, marital status, state of health, etc., which are statistically relevant to suicide. This provides a statistical-relevance explanation of her suicide.[25]

To say that the occurrence of an event of one type is statistically relevant to that of an event of another type is simply to say that the two are not statistically independent. In other words, statistical-relevance explanations on the first level explain individual occurrences on the basis of statistical dependencies. Statistical dependencies are improbable coincidences in the sense that dependent events occur in conjunction with a probability greater (or less in the case of negative relevance) than the product of their separate probabilities. Improbable coincidences demand explanation; hence, the statistical relevance relations invoked at the first level require explanation. The type of explanation required is, I believe, causal. Reichenbach (1956, § 19) formulated this thesis in his *principle of the common cause*. If all of the lights in an entire section of a city go out simultaneously, we explain this coincidence in terms of a power failure, not by the chance burning out of each of the bulbs at the same time. If two term papers are identical, and neither has been copied from the other, we postulate a common source (e.g. a paper in a fraternity file). Given similar patterns of rings in logs cut from two different trees, we explain the coincidence in terms of the rainfall in the area in which the two trees grew.

Given events of two types, A and B, which are positively relevant to one another, we hunt for a common cause C which is statistically relevant to both A and B.[26] C absorbs the dependency between A and B in the sense that the probability of A & B given C is equal to the product of the probability of A given C and the probability of B given C. The question naturally arises: Why should we prefer, for explanatory purposes, the relevance of C to A and C to B over the relevance of A to B which we had in the first place? The answer is that we can trace a spatio-temporally continuous causal connection from C to A and from C to B, while the relation between A and B cannot be accounted for by any such direct continuous causal relation. This is especially clear when A and B lie outside of one another's light cones.[27]

Improbable coincidences may have common effects as well as common causes, but their common effects do not explain the coincidences. Suppose that the only two ambulances in a town collide as they converge upon the scene of a serious automobile accident to which they had been summoned. The coincidence of their meeting is explained in terms of messages sent from a common source calling them to a particular place.[28] Suppose they were called to an accident in which the occupants of an automobile were seriously injured when it crashed into a truck at high speed. Suppose further that the people in the automobile were fleeing from the scene of a dastardly crime, and that, as a result of the collision between the two ambulances, the criminals died because they could not be taken to the hospital for treatment. We would not explain the collision of the ambulances as a case of justice prevailing in the 'punishment' of the criminals. This is not mere prejudice against teleological explanations; it results from the *fact* that the probability of the collision of the ambulances is not affected by the life or death, just or unjust, of any victim of the crash. It seems to be a basic and pervasive feature of the macrocosm that common causes, prior in time, can absorb the statistical relevance relations in improbable coincidences, while common effects (subsequent in time) cannot absorb these relevance relations.[29] Explanations thus exhibit a temporal asymmetry which is quite distinct from that of inferences.

I should like to close by offering a rough, but general, characterization of scientific explanation, followed by a challenge to which it gives rise. It seems to me that the nature of scientific explanation can be summed up as follows:

> To give scientific explanations is to show how events and statistical regularities fit into the causal network of the world.[30]

If this cannot be taken as a thesis supported by example and argument, it can, I believe, be advanced as a reasonable conjecture. It gives rise, however, to one of the most serious problems in current philosophy of science, namely, to provide an explication of causality without violating Hume's strictures against hidden powers and necessary connections. That we need such a characterization of causality, regardless of our attitude toward the role of causality in scientific explanation, is evident from the fundamental role played by causal relations in the basic space-time structure of the physical world. Since we need such an explication anyhow, the fact that our treatment of scientific explanation involves causal relations is no ground for objection to it.

University of Arizona, Tucson

NOTES

* I should like to express my gratitude to the National Science Foundation for support of research on scientific explanation and related topics.
[1] The classic article is Hempel *et al.* (1948).
[2] A serious problem about the nature of inductive 'inferences' or 'arguments' arises because of Carnap's denial of the existence of "rules of acceptance" in his system of inductive logic. I have discussed this issue in some detail in 'Hempel's Conception of

Inductive Inference in Inductive-Statistical Explanation' (forthcoming). In this essay I am construing 'argument' in the usual sense of a logical structure with premises and conclusions, governed by some sort of 'rule of acceptance'. Hempel's writings have conveyed to me, as well as to many others, I believe, the impression that he construes the term in this same way in his discussions of inductive-statistical explanation. In any case, whether Hempel construes explanations as arguments in this straightforward sense or not, there is no shortage of other philosophers who do.

[3] Neither Hempel nor anyone else, I suppose, has ever maintained that every sound argument is an explanation, and to attack such a thesis would certainly be to attack a straw man. In order to qualify as explanations, arguments must fulfill a number of conditions, and these have been carefully spelled out. Hempel, as well as many others, have claimed that every scientific explanation is an argument. It is this latter thesis that I am attempting to call into question.

[4] These examples, and many others like them, can be schematized so as to fulfill all of Hempel's requirements for deductive-nomological (D-N) explanations. This is shown in detail in Section 2 of Salmon (1970).

[5] This requirement is explicitly formulated and discussed in Section 45B of Carnap (1950).

[6] If '$c(h, e)$' designates the inductive probability or degree of confirmation of hypothesis h on evidence e, then i is irrelevant to h in the presence of e if and only if

$$c(h, e \cdot i) = c(h, e).$$

[7] If '$P(A, B)$' denotes the statistical probability of attribute B in reference class A, then C is statistically irrelevant to the occurrence of B within A if and only if

$$P(A \cap C, B) = P(A, B).$$

Note that this definition of statistical irrelevance is formally identical to the definition of irrelevance of evidence in the preceding note.

[8] Examples of this sort are discussed in Section 2 of Salmon (1970) where they are shown to conform to Hempel's requirements for inductive-statistical (I-S) explanations.

[9] I discuss this example in Section 8 of Salmon (1970) and there provide references to a number of other discussions of it.

[10] See, for example, the Introduction to Smith (1958). The point is illustrated by remarks on the edibility of certain species:
#11 (p. 34), helvella infula, "Poisonous to some, but edible for most people. Not recommended."
#87 (p. 126), cantharellus floccosus, "Edible for some people and NOT for others. "
#31 (p. 185), chlorophyllum molybdites, "Poisonous to some but not to others. Those who are not made ill by it consider it a fine mushroom. The others suffer acutely."

[11] Hempel was fully aware of this problem, and he discussed it explicitly (1965, pp. 410–412).

[12] It seems to me that Baruch Brody (1975, p. 71) missed this point when he wrote:

It should be noted that there are some cases of statistical explanation where the explanans does provide a high enough degree of probability for the explanandum, so Hempel's requirements laid down in his inductive-statistical model are satisfied, but does not differentiate between the explanadum and some of its alternatives. Thus, one can explain, even according to Hempel, the die coming upon one in 164 of 996 throws by reference to the fact that it was a fair die tossed in an unbiased fashion; such a die has, after all, a reasonably high probability of coming upon one in 164 out of 496 (sic!) throws. But the same explanans would also explain its coming upon one in 168 out of 996 throws. So it doesn't even follow from the fact than an explanation meets all of Hempel's requirements for statistical explanations that it meets the requirement that an explanans must differentiate between the explanandum and its alternatives.

The fact is that a fair die, tossed 996 times in an unbiased fashion has a probability of about 0.0336 of showing side one in 164 throws. By no stretch of the imagination can this be taken as a case in which Hempel's high probability requirement is satisfied. The probability that in the same number of throws with the same die the side one will show 168 times is nearly the same, about 0.0333. As Hempel has observed, if two events are incompatible, they cannot both have probabilities that are over 0.5 – a minimal value, I should think, for any probability to qualify as 'high'.

[13] For a fuller discussion of homogeneity, including a quantitative measure of degree of homogenity, see Salmon (1970), Section 6. See Greeno (1970) for an information-theoretic treatment of these concepts, especially the definition of 'information transmitted'. Both are reprinted in Salmon et al. (1971).

[14] See Salmon (1970) Section 9 for fuller discussion of this claim.

[15] Hempel (1965, pp. 381–403) offers recovery from a streptococcus infection upon treatment by penicillin as an example of inductive-statistical explanation. I believe it is now possible in principle to provide deductive-nomological explanations of such cures, for the chemistry of bacterial resistance to penicillin seems now to be understood (Cohen, 1975).

[16] I have discussed this issue briefly in the 'Postscript 1971' (Salmon et al., p. 106), indicating the difficulties that remain in A. Wald's refinement of von Mises's definition. The basic tool for overcoming these problems was provided by Church (1940). Subsequent work on randomness has not reduced the value of Church's fundamental contribution in the context of the present discussion.

[17] By 'non-trivial' I mean homogeneous references classes A in which some, but not all, elements have the attribute B.

[18] In personal conversation, Professor Hempel has expressed what seem to me to be reservations concerning the necessity of relativization of inductive-statistical explanations to knowledge situations in every instance, but I have not found such qualifications in his published writings. The question of the essentiality of epistemic relativization involves subtle issues whose detailed discussion must be reserved for another occasion. In this paper I shall confine my efforts to the attempt to draw out the consequences of what I take to be Hempel's published view.

[19] This argument is elaborated more fully in Salmon (1974). In private conversation, I. Niiniluoto pointed out that in infinite reference classes it may be possible to construct infinite sequences of partitions that do not terminate in trivially homogeneous subclasses. It is clear that no such thing can happen in a finite reference class. This is, therefore, one of those important points at which the admitted idealization involved in the use of infinite probability sequences (reference classes) must be handled with care.

[20] The issue of temporal asymmetry is discussed at length, including such examples as Bromberger's flagpole, in Section 12 of Salmon (1970).

[21] Ten people were in fact killed in a tragic head-on collision in Arizona. A pick-up truck crossed the center line on a straight stretch of road with clear visibility, striking an on-coming car, with no other traffic present. Prediction of such an accident would have been out of the question.

[22] It is often possible to infer the nature of a cause from a partial effect, but it is normally impossible to infer the nature of an effect from knowledge of a partial cause.

[23] See note 16.

[24] See Hempel (1965, p. 250) for the 1948 statement, but see note 6 (added in 1964) on the same page. The later view is more fully elaborated in Hempel (1965, pp. 347–354).

[25] Some philosophers would object to calling such assemblages of probabilities 'explanations'. Some other term, such as 'statistical systematization' might be preferred. I am fairly sympathetic to this view, and have some inclination to believe that explanation in a fuller sense occurs only when we move to the next level.

[26] Unless, of course, we can find a *direct* causal dependency, such as one student copying the work of another.

[27] This type of causal explanation is discussed in considerable detail in Salmon (1975).

[28] In the normal course of things, they might have collided at a particular location without having received a common call. This presumably, would be even less probable than the type of collision that actually occurred.

[29] See Reichenbach (1956, § 19) for fuller discussion, especially his concept of a *conjunctive fork*.

[30] Causal relations, as I am conceiving of them in this context, need *not* be deterministic; they are, instead, a species of statistical relevance relations. See Salmon (1975) for fuller elaboration.

BIBLIOGRAPHY

Braithwaite, R. B.: 1953, *Scientific Explanation*, Cambridge University Press.

Brody, B.: 1975,'The Reduction of Teleological Sciences', *American Philosophical Quarterly* **12.**

Carnap, R.: 1950, *Logical Foundations of Probability*, University of Chicago Press, Chicago (2nd ed., 1962).

Church, A.: 1940, 'On the Concept of a Random Sequence', *Bulletin of the American Mathematical Society* **46,** 130–135.

Coffa, J. A.: 1974, 'Hempel's Ambiguity', *Synthese* **28,** 161–164.

Cohen, S. N.: 1975, 'The Manipulation of Genes', *Scientific American* **233,** 24–33.

Greeno, J. G.: 1970, 'Evaluation of Statistical Hypotheses Using Information Transmitted', *Philosophy of Science* **37,** 279–293. Reprinted in Salmon *et al.* (1971) under the title, 'Explanation and Information', pp. 89–104.

Hempel, C. G.: 1959, 'The Logic of Functional Analysis', in L. Gross (ed.), *Symposium on Sociological Theory*, Harper and Row, New York. Reprinted, with revisions, in Hempel (1965), pp. 297–330.

Hempel, C. G.: 1962a, 'Deductive Nomological vs Statistical Explanation', in H. Feigl and G. Maxwell (eds.), *Minnesota Studies in the Philosophy of Science* **III,** University of Minnesota Press, Minneapolis, pp. 98–169.

Hempel, C. G.: 1962b, 'Explanation in Science and in History', in R. Colodny (ed.), *Frontiers of Science and Philosophy*, University of Pittsburgh Press, Pittsburgh, pp. 7–33.

Hempel, C. G.: 1965, *Aspects of Scientific Explanation*, The Free Press, New York.

Hempel, C. G., and P. Oppenheim: 1948, 'Studies in the Logic of Explanation', *Philosophy of Science* **15**, 135–175. Reprinted in Hempel (1965), 245–295.

Jeffrey, R. C.: 1969, 'Statistical Explanation vs Statistical Inference', in N. Rescher (ed.), *Essays in Honor of Carl G. Hempel*, D. Reidel Publishing Co., Dordrecht and Boston, pp. 104–113. Reprinted in Salmon *et al.* (1971), pp. 19–28.

Kyburg, H. E., Jr.: 1970, 'Conjunctivitis', in M. Swain (ed.), *Induction, Acceptance, and Rational Belief*, D. Reidel Publishing Co., Dordrecht and Boston, pp. 55–82.

Nagel, E.: 1961, *The Structure of Science*, Harcourt, Brace & World, New York & Burlingame.

Popper, K. R.: 1959, *The Logic of Scientific Discovery*, Basic Books, New York. (English translation of 1935, *Logik der Forschung*, Julius Springer, Vienna.)

Quine, W. V.: 1951, 'Two Dogmas of Empiricism', *Philosophical Review* **60.**

Reichenbach, H.: 1956, *The Direction of Time*, University of California Press, Berkeley and Los Angeles.

Salmon, W. C.: 1970, 'Statistical Explanation', in R. Colodny (ed.), *The Nature and Function of Scientific Theories*, University of Pittsburgh Press, Pittsburgh, 173–231. Reprinted in Salmon *et al.* (1971), pp. 29–88.

Salmon, W. C.: 1974, 'Comments on "Hempel's Ambiguity" by J. A. Coffa', *Synthese* **28**, 165–169.

Salmon, W. C.: 1975, 'Theoretical Explanation', in S. Körner (ed.), *Explanation*, Basil Blackwell, Oxford.

Salmon, W. C., *et al.*: 1971, *Statistical Explanation and Statistical Relevance*, University of Pittsburgh Press, Pittsburgh.

Scriven, M.: 1959, 'Explanation and Prediction in Evolutionary Theory', *Science* **130.**

Smith, A. H.: 1958, *The Mushroom Hunter's Guide*, University of Michigan Press, Ann Arbor.

von Mises, R.: 1964, *Mathematical Theory of Probability and Statistics*, Academic Press, New York.

AMOS TVERSKY AND DANIEL KAHNEMAN

CAUSAL THINKING IN JUDGMENT UNDER UNCERTAINTY

Many of the decisions we make, in trivial as well as in crucial matters, depend on the apparent likelihood of events such as the keeping of a promise, the success of an enterprise, or the response to an action. In general, we do not have adequate formal models to compute the probabilities of such events. Consequently, most evaluations of likelihood are subjective and intuitive. The manner in which people evaluate evidence to assess probabilities has aroused much research interest in recent years, e.g. Edwards (1968), Slovic (1972), Slovic, Fischhoff, and Lichtenstein (1975), Kahneman and Tversky (1973), and Tversky and Kahneman (1974). This research has identified different heuristics of intuitive thinking and uncovered characteristic errors and biases associated with them. The present paper is concerned with the role of causal thinking in the evaluation of evidence and in the judgment of probability.

Students of modern decision theory are taught to interpret subjective probability as degree of belief, i.e. as a summary of one's state of information about an uncertain event. This concept does not coincide with the lay interpretation of probability. People generally think of the probability of an event as a measure of the propensity of some causal process to produce that event, rather than as a summary of their state of belief. The tendency to regard probabilities as properties of the external world rather than of our state of information characterizes much of our perception. We normally regard colors as properties of objects, not of our visual system, and we treat sounds as external rather than internal events. In a similar vein, people commonly interpret the assertion 'the probability of heads on the next toss of this coin is 1/2' as a statement about the propensity of the coin to show heads, rather than as a statement about our ignorance regarding the outcome of the next toss. The main exceptions to this tendency are assertions of complete ignorance, e.g. 'I haven't the faintest idea what will happen in the coming election'.

Butts and Hintikka (eds.), Basic Problems in Methodology and Linguistics, 167–190.
Copyright © 1977 by D. Reidel Publishing Company, Dordrecht-Holland. All Rights Reserved.

The interpretation of probability as propensity leads people to base their judgments of likelihood primarily on causal considerations, and to ignore information that does not have causal significance. The neglect of evidence that does not lend itself to causal interpretation produces characteristic errors in judgments of probability. The present paper investigates this phenomenon in three different contexts. In the first section, we show that the assessment of a conditional probability $P(A/B)$ is determined mainly by the perceived causal impact of B on A, even when this mode of judgment yields paradoxes and inconsistencies. The second section concerns the assessment of the posterior probability of an event, given the base-rate frequency of that event and some additional specific evidence. We show that base-rate information is generally ignored unless it is given a causal interpretation, and discuss the conditions under which such causal interpretations arise. In the third section, we study the assessment of the probability of a hypothesis H given two items of information D_1 and D_2, on the basis of the prior probability $P(H)$, and the conditional probabilities $P(H/D_1)$ and $P(H/D_2)$. We show that people adopt different rules of combination of evidence depending on whether the data D_1 and D_2 are given a causal interpretation.

1. INTERPRETATIONS AND MISINTERPRETATIONS OF CONDITIONAL PROBABILITY

The tendency to interpret probability statements in causal terms is nowhere more evident than in the context of conditional probability. Whenever two events A and B can be linked by a causal schema, the perceived causal relations between the events play a dominant role in the intuitive assessment of the conditional probabilities $P(A/B)$ and $P(B/A)$. For example, people tend to answer the question, 'What is the probability of a major economic crisis within a year of the next presidential election, given that Mr. X is elected?' in terms of the impact of Mr. X's presidency on the economy. The reliance on causal relations to assess conditional probabilities makes it quite difficult to answer questions in which the conditioning event occurs after the target event, e.g. 'What is the probability that Mr. X will have been elected, if there is a major economic crisis within a year of the next election?'. People find such questions somewhat bewildering and often respond by assessing $P(B/A)$ instead of the required $P(A/B)$, if A is the earlier of the two events. In the logic of information, of course, the temporal sequence per se is irrelevant since the occurrence of an event can provide as much information about events that precede it as about events that follow it.

Causal thinking, however, follows the time arrow. Specifically, when told to assume that a particular conditioning event has occurred, people are prone to focus on the causal impact of this event on the future and to neglect its significance as a source of information about the past. We now discuss a class of problems, originally introduced by Turoff (1972), in which this bias produces inconsistent and paradoxical probability assessments.

PROBLEM 1 (Turoff).   Let C be the event that within the next 5 years Congress will have passed a law to curb mercury pollution, and let D be the event that within the next 5 years, the number of deaths attributed to mercury poisoning in the U.S. will exceed 500. Let $\bar{C}$ and $\bar{D}$ denote the complements of C and D respectively.
Question: Which of the two conditional probabilities, $P(C/D)$ and $P(C/\bar{D})$, is higher?
Question: Which of the two conditional probabilities, $P(D/C)$ and $P(D/\bar{C})$, is higher?

The overwhelming majority of respondents state that Congress is more likely to pass a law restricting mercury pollution if the death toll exceeds 500 than if it does not, i.e. $P(C/D) > P(C/\bar{D})$. Most people also state that the death toll is less likely to reach 500 if a law is enacted within the next five years than if it is not, i.e. $P(D/C) < P(D/\bar{C})$. These judgments reflect the beliefs that a high death toll would increase the pressure to pass an anti-pollution measure, and that such a measure would be effective in the prevention of mercury poisoning. In a sample of 68, we found that 85% of respondents chose the modal answer to both questions. However, this seemingly plausible pattern of judgments violates the most elementary rules of conditional probability.

Clearly, $P(C/D) > P(C/\bar{D})$ implies $P(C/D) > P(C)$. Furthermore, the inequality

$$P(C/D) = \frac{P(C \text{ and } D)}{P(D)} > P(C)$$

holds if and only if $P(C \text{ and } D) > P(C) P(D)$ which holds if and only if

$$P(D) < \frac{P(C \text{ and } D)}{P(C)} = P(D/C)$$

which in turn implies $P(D/C) > P(D/\bar{C})$, provided $P(C)$ and $P(D)$ are non-zero. Hence, $P(C/D) > P(C/\bar{D})$ implies $P(D/C) > P(D/\bar{C})$ contrary to the prevailing pattern of judgment.

It is easy to construct additional examples of the same type in which people's intuitions violate the probability calculus. Such examples consist of a pair of events A, B, such that the occurrence of A increases the likelihood of the subsequent occurrence of B, while the occurrence of B decreases the likelihood of the subsequent occurrence of A. For example, consider the following problem.

PROBLEM 2.   Let A be the event that before the end of next year, Peter will have installed a burglar alarm system in his home. Let B denote the event that Peter's home will be burglarized before the end of next year.
Question: Which of the two conditional probabilities, $P(A/B)$ and $P(A/\bar{B})$, is higher?
Question: Which of the two conditional probabilities, $P(B/A)$ and $P(B/\bar{A})$, is higher?

A large majority (56 out of 68) believe both that $P(A/B) > P(A/\bar{B})$ and $P(B/A) < P(B/\bar{A})$, contrary to the laws of probability. The reasoning behind these judgments is presumably causal: the presence of a burglar alarm system reduces the chances of a burglary, while the occurrence of a burglary motivates the victim to install a burglar alarm system.

These examples illustrate the profound contrast between the informational analysis of evidence embodied in probability theory and the causal interpretation of evidence that people naturally adopt. To appreciate the nature of this contrast, let us analyze the structure of Problem 2.

First consider $P(A/B)$, the conditional probability that Peter will install an alarm-system in his home before the end of next year, assuming that his home will be burglarized some time during this period. The alarm system could be installed either before or after the burglary. The information conveyed by the condition, i.e. the assumption of a burglary, has causal significance with respect to the future and diagnostic significance with respect to the past. Specifically, the occurrence of a burglary provides a cause for the subsequent installation of an alarm system, and it provides a diagnostic indication that the house had not been equipped with an alarm system at the time of the burglary. Thus, the causal impact of the burglary increases the likelihood of the alarm system while the diagnostic impact of the burglary decreases this likelihood.

Formally, let $A(t)$ denote the event that A (i.e. the installation of an alarm system) occurs at time t, and let B_t and B^t denote, respectively, the events that B (i.e. the burglary) occurs before and after t. Clearly, $P(B_t/B) + P(B^t/B) = 1$, and

$$P(A(t)/B) = P(A(t)/B_t)P(B_t/B) + P(A(t)/B^t)P(B^t/B).$$

That is, the conditional probability of A at time t given B (i.e. $P(A(t)/B)$) is a weighted average of the conditional probability of $A(t)$ given that B occurs before t (i.e. $P(A(t)/B_t)$) and the conditional probability of $A(t)$ given that B occurs after t (i.e. $P(A(t)/B^t)$). The conditional probability of A given B, therefore, is obtained by integration (or summation) over time. Thus,

$$P(A/B) = \int P(A(t)/B) \, dt$$

$$= \int (P(A(t)/B_t)P(B_t/B) + P(A(t)/B^t)P(B^t/B)) \, dt$$

and

$$P(A/\bar{B}) = \int P(A(t)/\bar{B}) \, dt.$$

Since a burglary induces people to install a burglar alarm system, and the installation of such a system reduced the likelihood of a subsequent burglary,

$$P(A(t)/B_t) > P(A(t)/\bar{B}) > P(A(t)/B').$$

Note that the causal impact of B on $A(t)$ is incorporated in the left-hand probability (conditioned on B_t) while the diagnostic impact of B on $A(t)$ is incorporated in the right-hand probability (conditioned on B'). The conditional probability of $A(t)$ given no burglary $P(A(t)/\bar{B})$ lies between these values. In order to decide whether $P(A/B)$ exceeds $P(A/\bar{B})$, therefore, the *causal* component $P(A(t)/B_t)$ and the *diagnostic* component $P(A(t)/B')$ should be properly weighted. The nearly unanimous judgment that $P(A/B) > P(A/\bar{B})$ indicates that the causal impact of B dominates its diagnostic impact.

Precisely the same analysis applies to $P(B/A)$ – the probability that Peter's house will be burglarized before the end of next year, given that he will have installed an alarm system sometime during this period. Naturally, the presence of an alarm system is causally effective in reducing the likelihood of a subsequent burglary; it also provides a diagnostic indication that the occurrence of a burglary may have prompted Peter to install the alarm system. The causal impact of the alarm system reduces the likelihood of a burglary; the diagnostic impact of the alarm system increases this likelihood. That is,

$$P(B(t)/A_t) < P(B(t)/\bar{A}) < P(B(t)/A'),$$

where

$$P(B/A) = \int P(B(t)/A)\, \mathrm{d}t$$

$$= \int \left(P(B(t)/A_t)P(A_t/A) + P(B(t)/A')P(A'/A) \right) \mathrm{d}t$$

and

$$P(B/\bar{A}) = \int P(B(t)/\bar{A})\, \mathrm{d}t.$$

Here again, the prevalence of the judgment that $P(B/A) < P(B/\bar{A})$ indicates that the causal impact of A, reflected in $P(B(t)/A_t)$, dominates its diagnostic impact, reflected in $P(B(t)/A')$. Instead of weighing the causal and the diagnostic impacts of the evidence, people apparently identify the conditional probabilities $P(A/B)$ and $P(B/A)$ with their causal components. As was shown earlier, this causal interpretation of conditional probability leads to inconsistency.

The preceding argument demonstrates a basic incompatibility between the intuitions of intelligent adults and the logic of conditional probability. Indeed, Turoff (1972) argued from his example that judgments of conditional probabilities which are inconsistent with the formal calculus should nevertheless be accepted, and interpreted as 'causal probabilities'. Although Turoff's paradox arises under very special circumstances, it illustrates a phenomenon of much broader significance, i.e. the tendency to focus on the causal impact of evidence and to ignore considerations that do not have direct causal implications.

II.   ON THE USE AND NEGLECT OF BASE-RATE DATA

When people judge the probability of an event they usually have access to information of two types: specific and distributional. Specific information, or case data, consists of evidence about the particular case under consideration. Distributional information, or base-rate data, consists of knowledge about the relative frequency of that event in a relevant population. For example, the presenting symptoms of a patient or the results of a laboratory test provide specific information about the presence of a particular disease, while the base-rate frequency of that disease provides distributional information. Note that the present concept of distributional information does not coincide with the Bayesian concept of prior probability. The former is defined by the nature of the data, whereas the latter is defined by the temporal sequence of information acquisition.

One of the most dramatic violations of the rules of probability in intuitive judgments is that intuitive assessments of probability are determined exclusively by specific evidence, with little or no regard for distributional data. To illustrate, consider the following problem (Kahneman and Tversky, 1973).

PROBLEM 3.   A panel of psychologists interviewed a sample of 30 engineers and 70 lawyers, and summarized their impressions in thumbnail descriptions of those individuals. The following description has been drawn at random from the sample of 30 engineers and 70 lawyers.

"John is a 39 year old man. He is married and has two children. He is active in local politics. The hobby that he most enjoys is rare book collection. He is competitive, argumentative, and articulate."

Question: What is the probability that John is a lawyer rather than an engineer?

Problem 3 was given to a group of 85 respondents. Another group of 86 respondents answered another version of the same problem, the only difference being that the sample from which John was allegedly drawn was said to consist of 70 engineers and 30 lawyers. The median answer to Problem 3 was 0.95 in both groups. Thus, the base-rate frequency of engineers and lawyers had no effect on subjects' estimates. It follows readily from Bayes' rule that the probability of John being a lawyer should be considerably higher when the base-rate of lawyers is high than when it is low. Furthermore, the ratio of the posterior odds of the two groups should be $(7/3)^2$, see Kahneman and Tversky (1973).

In that paper, we related the neglect of base-rate to the hypothesis that people evaluate the probability that John is a lawyer rather than an engineer by the degree to which he is more representative of the stereotype of lawyers than of that of engineers. Since the similarity of a thumbnail description to the stereotype of a category is unaffected by the base-rate frequency of the category, a judgment of probability that is based exclusively on similarity or representativeness will be independent of base-rate frequency. While judgments of probability by representativeness are inherently insensitive to base-rate frequencies, the neglect of base-rate information is a very general phenomenon, which occurs even when considerations of representativeness or similarity are not involved in the assessment of probability. Indeed, we propose now the more general hypothesis that in the presence of specific evidence, base-rate data will be essentially ignored, except when they can be incorporated in a causal model which applies to the case under consideration. We first discuss some elementary demonstrations of this hypothesis in problems where both base-rate and case data are provided in numerical form.

PROBLEM 4.   A cab was involved in a hit-and-run accident at night. Two cab companies, the Green and the Blue, operate in that city. You are given the following data:
   (i) 85% of the cabs in the city are Green and 15% are Blue.
   (ii) A witness identified the cab as a Blue cab. The court tested his

ability to identify cabs under the appropriate visibility conditions. When presented with a sample of cabs (half of which were Blue and half of which were Green) the witness made correct identifications in 80% of the cases and erred in 20% of the cases.

Question: What is the probability that the cab involved in the accident was Blue rather than Green?

Several hundred subjects have been given this question. The most frequent answer is 80%, which is also the median of all answers. Thus, the intuitive judgment of probability coincides with the credibility of the witness and ignores the relevant base-rate, i.e. the relative frequency of Green and Blue cabs. It is instructive to contrast people's answers with the formal solution of the problem. To obtain the correct answer, let B and G denote respectively, the hypothesis that the cab involved in the accident was Blue or Green, and let W be the witness's report. By Bayes' rule in odds form:

$$\frac{P(B/W)}{P(G/W)} = \frac{P(W/B)P(B)}{P(W/G)P(G)} = \frac{(0.8)(0.15)}{(0.2)(0.85)} = \frac{12}{17}$$

and hence

$$P(B/W) = \frac{12}{12+17} = 0.41.$$

In spite of the witness's report, therefore, the hit-and-run cab is less likely to be Blue than Green, because the base-rate is more extreme than the witness is credible. The overwhelming majority of subjects, however, fail altogether to take the base-rate into account.

Base-rate information, however, is properly used in the absence of case data. When item (ii) is omitted from Problem 4, almost all subjects give the correct answer that the probability of the cab being Blue is 0.15. In the absence of case data the question is naturally viewed as a sampling problem, and the accident serves to define a sampling trial in which a single cab is drawn from the population of cabs in the city. As soon as pertinent case data about the hit-and-run cab is introduced, the base-rate no longer seems relevant. Apparently, people are unable to relate the color of the hit-and-run cab simultaneously to two different types of processes: random sampling of cabs, and imperfect indentification of color by a witness.

The neglect of base-rate data is a highly robust effect, which has been confirmed in a variety of contexts ranging from simple formal questions such as Problem 4 (Kahneman and Tversky, 1973; Hammerton, 1973; Lyon and Slovic, 1976) to complex realistic problems (Nisbett and Borgida, 1975; Borgida and Nisbett, 1976). In a recent doctoral dissertation, Maya Bar-Hillel (1975) has thoroughly investigated the neglect of base-rate using a wide variety of formal problems. Her results, and those of Lyon and Slovic (1976), show that base-rate data are ignored regardless of whether they are presented before or after the case data, and regardless of whether the information is given in numbers or in words, e.g. 'the great majority of cabs are green', 'the witness was correct in most of the cases'. Furthermore, base-rate data are ignored regardless of whether they confirm or disconfirm the relevant case data, e.g. the witness' identification. There was a small effect of extreme base-rate (e.g. 1:99), although the discrepancy between estimates and correct values was actually most extreme in these cases.

Bar-Hillel (1975) also explored the conditions under which base-rate data are not neglected, and constructed several problems in which they are combined with specific evidence. In order to understand when and why base-rate data are ignored, consider the following variation of Problem 4.

PROBLEM 5.   A cab was involved in a hit-and-run accident at night. Two cab companies, the Green and the Blue, operate in that city. You are given the following data:

(i) Although the two companies are roughly equal in size, 85% of cab accidents in the city involve Green cabs, and 15% involve Blue cabs.

(ii) As in Problem 4 above.
Question: What is the probability that the cab involved in the accident was Blue rather than Green?

Problem 5 is structurally identical to Problem 4, but the responses to the two problems were radically different. The median answer of a group of 72 subjects presented with Problem 5 was 0.55, which is not too far from the correct answer of 0.41. It should be pointed out that the answers of different subjects to Problem 5 varied widely, indicating a lack of consensus in the weighting of the base-rate and the case data. Nevertheless, the base-rate in this problem was no longer ignored. Note that Problem 5 differs from Problem 4 only in that the base-rate refers to the frequency of accidents rather than to the frequency of cabs According to our analysis, however, this is a crucial change. The difference between the base-rate frequencies of Blue and Green cabs in the city is viewed as an incidental statistical fact, which does not elicit a causal explanation, and does not make any specific Green cab more likely to be involved in an accident than any specific Blue cab. On the other hand, the difference between the base-rates of accidents for the two companies immediately elicits a causal explanation, i.e. that the drivers of the Green cabs are more reckless and less competent than the drivers of the Blue cabs. This explanation applies to any specific Blue and Green cab and increases the perceived likelihood that a Green rather than a Blue cab was involved in the accident.

More generally, we suggest that base-rate data will be combined with case data when the difference in the base-rate of outcomes is interpreted as a difference in the propensities to produce these outcomes. In Problem 4, the difference in the frequencies of Green and Blue cabs cannot be related to the propensity of any particular cab to be involved in an accident, and it is therefore ignored. In Problem 5, on the other hand, the difference in the frequency of accidents is interpreted as a difference in accident-proneness and it is therefore utilized.

The information provided in Problems 4 and 5 consists of base-rate and case data, i.e. the credibility of the witness. In the following problem, based on Bar-Hillel (1975), no specific evidence is given and the information is entirely distributional. Nevertheless, the results show that base-rate information which is interpreted as a propensity dominates other base-rate information which cannot be interpreted in this manner.

PROBLEM 6 (Bar-Hillel).   Consider the following hypotheses regarding suicide. In a population of young adults, 80% of the individuals are married and 20% are single. The percentage of deaths by suicide (relative to all other forms of non-natural deaths) is 15% among single individuals and 5% among married individuals.

Question: What is the probability that an individual, selected at random from those that had committed suicide, was single?

The correct answer to this problem follows readily from Bayes' rule. Let D denote death by suicide, and let M and S denote, respectively, married and single. Hence

$$\frac{P(S/D)}{P(M/D)} = \frac{P(D/S)}{P(D/M)} \frac{P(S)}{P(M)} = \frac{(0.15)(0.20)}{(0.05)(0.80)} = \frac{3}{4}$$

and $P(S/D) = 3/7 = 0.43$. The median and modal answer of a group of 73 subjects was 0.75. This value corresponds exactly to the ratio of suicide rates among single and married people, and reflects total neglect of the base-rate frequencies of these categories.

This result supports our analysis regarding the role of causal interpretation. Two items of information are supplied in this question: the base-rate frequencies of married and single individuals, and the ratio of suicide-rates in these categories. The second of these items is readily related to a causal schema of suicide, while the former is not. The differential suicide-rates imply that a single person has a stronger propensity to commit suicide than a married person. In contrast, the different frequencies of single and married individuals have no bearing on the causation of suicide. In accord with our hypothesis, base-rate information which is not linked to a causal schema is ignored in the presence of causally relevant evidence. This result does not depend on prior beliefs regarding suicidal tendencies in the population. When respondents were told that the suicide-rate is higher among married than among single people, the base-rate frequencies of these categories were again ignored (Bar-Hillel, 1975). If the above account is correct, the base-rate should not be ignored when it is causally related to the target outcome, as in the following problem.

PROBLEM 7.   Consider the following hypotheses regarding suicide. In a population of adolescents, 80% of suicide attempts are made by girls, and 20% by boys. The percentage of suicide attempts that result in death is 15% among boys and 5% among girls.

Question: What is the probability than an adolescent, selected at random from those who had died by suicide, was a girl?

Problem 7 is structurally identical to Problem 6. The classes of boys and girls correspond to the classes of single and married people, and the percentages of fatal suicide attempts among boys and girls correspond to the suicide rates among single and married people. The answers to these two equivalent problems, however, were quite different. The median answer to Problem 7 was 0.50, indicating that the base-rate of suicide attempts was not ignored. As in Problem 5, however, there was considerable variability in the estimates of 68 subjects, which ranged from 0.20 to 0.80.

Why is the base-rate frequency ignored in Problem 6 but not in Problem 7? According to the present analysis, the base-rate is properly utilized only when it can be interpreted as a propensity that is causally related to the target outcome. The base-rate frequency of single and married individuals in the relevant population does not affect in any way the propensity of any particular individual to commit suicide. In contrast, the differential base-rate of suicide attempts among boys and girls is readily interpreted to mean that adolescent girls are more prone to attempt suicide than adolescent boys. In addition, the subject is informed in Problem 7 that girls have a lower propensity to die from a suicide attempt. Since both items of information are readily interpreted causally, neither item dominates the other. In Problem 6, in contrast, one item (the differential suicide rate) has a causal character, while the other (the relative frequency of single and married individuals) is not causally related to the target outcome. Consequently, the former dominated the latter completely.

The above results suggest the presence of different levels of perceived relevance of data. When a problem includes several items of information, people combine items which belong to the same level of perceived relevance. When a problem includes items of high and low perceived relevance, the latter are ignored. The perceived relevance of case data is generally high. The perceived relevance of base-rate data is high when the data are interpreted as differential propensities to produce the target outcome. When no such interpretation is available, the perceived relevance of base-rate is low.

Another manner in which base-rate data acquire high perceived relevance is illustrated in the following problem that was presented in two different versions to two separate groups of 47 subjects.

PROBLEM 8.   Among high school seniors in a given school, 5% (or 40%) were awarded scholarships.

David is a senior in that school who was described by his counselor as "industrious, intelligent, and responsible, does well above average in structured tasks, but lacks intellectual curiosity".
Question: What is the probability that David received a scholarship?

The answer to Problem 8 is sensitive to the base-rate frequency of recipients of the scholarship. The median estimate was 0.25 when the base-rate was 5%, and 0.50 when the base-rate was changed to 40%. The change of base-rate, in this problem, alters the perception of the scholarship from one that is hard to get to one that is easy to get, and the estimates of David's chances vary accordingly. Thus, base-rate information can be used to define the outcome as easy or hard to produce, when the judge has no prior conception of the outcome, as is the case with the scholarship.

There is evidence, however, that people do not readily alter their prior conception of outcomes when exposed to relevant distributional data. Impressive demonstrations of this hypothesis have been reported by Nisbett and Borgida (1975). They presented subjects with detailed information about the procedure and results of a well-known experimental investigation of people's willingness to help a stranger in distress (Darley and Latané, 1968). In that experiment, six men (of whom one was a confederate of the experimenter) participated in a discussion of personal problems. Under the pretext of maintaining privacy, the participants were seated in separate booths and the conversation was held through an intercom system. The confederate, who was the first to speak, mentioned that he was prone to seizures. Then he began to stammer, indicated that one of his seizures was coming on, and asked for help. His last comments were "I'm gonna die-- er-- er. I'm, ... , die--er--er--seizure--er-- (choking sounds, then silence)".

The object of the helping experiment was to find out what proportion of participants would try to help, and how soon they would do so. The surprising result was that not one of the fifteen subjects rushed to help the confederate when he asked for help. Four subjects came out of their booths by the end of the speech when the man was choking, and six never came out of their booths at all. Nisbett and Borgida presented these data to their subjects in order to test their effect on predictions of the behavior of specific individuals. The subjects were shown brief film clips of interviews, allegedly conducted with three of the participants in the helping experiment. The interviews were not directly pertinent to helping behavior but they provided a general impression of the participant's personality. The subjects were then asked to guess how each of the three participants had behaved in the helping experiment.

Nisbett and Borgida found that knowledge of the distribution of helping behavior in the original experiment did not affect predictions of specific cases. Subjects who knew the results of the helping experiment and subjects who were not given this information made identical predictions concerning the helping behavior of the men shown in the film. Since these men appeared normal and friendly in the films, informed and uninformed subjects alike tended to guess that they had tried to help the stranger. The knowledge that helping had been infrequent did not affect the predictions. Nisbett and Borgida comment "It is interesting to speculate just what kind of monstrous target case these subjects would have had to witness before they would guess that he would behave in a fashion that they knew to be modal" (p. 940).

The results of the helping experiment should have led the subjects of Nisbett and Borgida to realize that rushing out of one's booth to help a stricken stranger is much harder than it appears to be. The predictions of helping behavior should have been modified accordingly, as in the scholarship example. This did not occur, and the exposure to base-rate evidently did not lead the subjects to revise their views of the difficulty of helping.

Nisbett and Borgida (1975) conducted another study in which the subjects were given a detailed description of the procedure of the helping experiment but were not given the results. Instead, they were shown film interviews with two alleged participants, and were told afterwards that the individuals shown in the films had remained in their booths and had not come to the stranger's help. Following this exposure to two individual cases, subjects were asked to predict the general results of the helping experiment. This information greatly reduced the estimated base-rate of helping behavior. Having seen two seemingly normal individuals who did not help induced a general expectation that others would also fail to help. Thus, estimates of the base-rate frequency of helping were readily altered by exposure to individual cases, while predictions of the helping behavior of specific individuals were unaffected by exposure to base-rate information. Nisbett and Borgida conclude that "people's unwillingness to deduce the particular from the general is matched only by their willingness to infer the general from the particular" (p. 939).

Let us now review our conclusions concerning the uses and misuses of base-rate information. Base-rate knowledge

(a)   is appropriately used in the absence of specific evidence;

(b)   is generally neglected in the presence of specific evidence (Problems 3,4,6);

(c)   is combined with specific evidence when the judge forms a causal model which 'explains' the base-rate and applies to the case in question (Problems 5,7);

(d)   is used to define an outcome as 'easy' or 'hard' to produce or obtain, when the judge has no prior conception of the outcome scale (e.g. the scholarship example);

(e)   is not used to revise an outcome scale with which the judge believes himself to be familiar (e.g. the scale of helping behavior);

(f)   is readily altered by exposure to a few specific instances.

Nisbett and Borgida (1975; Borgida and Nisbett, 1976) explained the neglect of base-rate and the over-eager reliance on specific instances in terms of the relative vividness of the two types of information. They point out that case data are typically concrete and vivid, whereas distributional data are typically abstract and pallid. While vividness is undoubtedly an important factor in the weighting of evidence, we propose that the range of phenomena associated with the use and neglect of base-rate knowledge is best understood in terms of the prevalent tendency to think causally.

To judge the probability of an event, people commonly relate it to a specific causal schema, in which that event serves the role of consequence or cause. Distributional information is perceived as relevant only when it is incorporated into a causal schema which pertains to the case under study. Base-rate evidence can be incorporated into a causal schema in two ways: when it induces differential propensities ·to produce the target outcome, as in Problems 5 and 7; or when it defines the outcome scale, as in the case of the scholarship in Problem 8. Base-rate evidence that is not interpreted in either of these two modes is treated as an incidental statistic, which is ignored in the presence of causally relevant evidence.

This analysis emphasizes the distinction between distributional data which induces propensities for specific outcomes, and distributional data which cannot be used in this fashion. The critical difference between the base-rate frequency of cabs and the base-rate frequency of accidents is not in vividness. The latter appears more relevant than the former because it suggests a causal interpretation in terms of recklessness, which is pertinent to the individual accident. In contrast, the base-rate frequency of cabs in the city does not affect the propensity of any particular cab to be involved in accidents.

Our analysis suggests that base-rate data must be *explained*, if they are to become relevant to the prediction of specific cases. It will come as no surprise to teachers that people often accept statistical data passively, without any attempt to generate explanations for them. The neglect of base-rate data is all the more striking in the light of the great impact of individual examples, demonstrated by Nisbett and Borgida (1975). We suggest that this difference reflects the relative effectiveness of the two types of information in bringing about a change in a causal model. The exposure to a seemingly normal and friendly man, with the information that he did not help a stranger in distress, violates one's prior conception. This discrepancy can only be resolved by modifying one's conceptions and expectations of helping behavior. The distributional data for helping behavior also violate prior expectations, but since they are often treated as mere statistics, they are less effective in initiating the process by which conceptions of behavior are modified.

In discussing the persuasive impact of specific instances, Nisbett and Borgida (1975) suggest the disconcerting conclusion that exposure to vivid and concrete examples may be the only way for people to fully assimilate statistical generalizations. This proposition has profound implications for education in general, and for the teaching of social science in particular. We do not wish to deny the effectiveness of concrete examples. Our analysis, however, leads to a somewhat less pessimistic view of people's ability to acquire new and usable knowledge through exposure to statistical data. Specifically, it appears that distributional information or base-rate data become effective when the data are explained or incorporated into an appropriate causal schema. We speculate that novel distributional information, e.g. about helping behavior, could be made effective if the recipients of this information were required to explain it. Encouraging the learner to search for explanations of statistical data and not to accept them passively, may be a viable alternative to the teaching of generalizations through specific examples.

III. AGGREGATION AND MISAGGREGATION OF PROBABILISTIC DATA

In this section we deal with problems of the following type.

PROBLEM 9.   Consider the following hypotheses concerning the causes of death.

(i) The chance of death from heart failure is 5% among males.

(ii) The chance of death from heart failure is 10% among males who are heavy smokers.

(iii) The chance of deaths from heart failure is 45% among males with congenital high blood pressure.

Dick is a heavy smoker with congenital high blood pressure.
Question: What is the probability that Dick will die of heart failure?

In this class of problems, one is given three items of information: the base-rate frequency of the target outcome, e.g. the frequency of heart failure in the male population $P(H)$, and the conditional probabilities of the target outcome given two data, $P(H/D_1)$ and $P(H/D_2)$. One is asked to assess $P(H/D_1$ and $D_2)$ on the basis of these data. The answer to such problems cannot be computed without additional assumptions concerning the dependency between D_1 and D_2. Indeed, it is possible to construct examples in which both D_1 and D_2 support the hypothesis

H while their conjunction, and even their disjunction, provide evidence against it (see Salmon, 1975).

Recall that in Problem 9 each datum alone increases the likelihood of the target outcome, i.e. $P(H/D_i) > P(H)$, $i = 1$, 2. In the problems studied in this section, it is reasonable to assume in addition that the following incremental property is satisfied:

$$P(H/D_1 \text{ and } D_2) > P(H/D_i), \qquad i = 1, 2.$$

That is, H is more likely given both D_1 and D_2 than it is on the basis of either D_1 or D_2 alone.

The incremental property is compatible with the common causal intuition according to which each additional cause increases the likelihood of the outcome. In Problem 9, for example, both heavy smoking and congenital high blood pressure are naturally viewed as causal factors that increase the likelihood of heart failure. Indeed, a significant majority of respondents (44 out of 62) satisfied the incremental property, and the median estimate was 0.50. Thus, the intuitive answers to Problem 9 were in qualitative agreement with the normative solution. As the following problem demonstrates, however, a different rule of combination of evidence is applied when the data are not naturally interpreted in causal terms.

PROBLEM 10.   Bill has been referred by his physician to the hospital with suspicion of a malignant tumor. Following the examination the following data were obtained.

(i) The chance of malignant tumor is 5% among patients referred to the hospital for such examinations.

(ii) The hematologist who examined Bill's blood test estimated the chance of a malignancy to be 10%.

(iii) The radiologist who examined Bill's X-ray estimated the chance of a malignancy to be 45%.
Question: What is the probability that Bill has a malignant tumor?

Problem 10 is structurally similar to Problem 9. In both problems each datum provides support for the hypothesis in question, and it appears reasonable to assume the incremental property.[1] The critical difference between the problems is that the evidence in Problem 9 (i.e. heavy smoking and congenital high blood-pressure) provides causes for heart failure, whereas the evidence in Problem 10 (i.e. the judgments of experts) provides estimates of the likelihood of malignancy rather than causes for its presence. If each of the two experts provides independent support for the hypothesis of a malignancy, then one's degree of belief in that hypothesis should be higher when given both estimates than when given each estimate alone. But since probability is generally interpreted as a property of a hypothesis rather than of a state of belief, people are expected in this case to apply the logic of estimation and average the estimates in the same way that they average estimates of height or life expectancy. (We use averaging to refer to a general weighted average, not necessarily an arithmetic mean.)

The responses to Problem 10 confirmed this hypothesis. A significant majority of the subjects (38 out of 50) violated the incremental property, and the median estimate was 0.35. Apparently, $P(H/D_1)$ and $P(H/D_2)$ are treated as independent estimates of the probability of a malignancy, and hence they are averaged. The tendency to average rather than compound probability estimates has been observed in other problems of the same general type (see, e.g. Bar-Hillel, 1975).

Note than in the normative theory, given $P(H)$, $P(H/D_1)$, and $P(H/D_2)$, the value of $P(H/D_1$ and $D_2)$ is determined only by the degree of dependency between D_1 and D_2. If the estimates of the two experts in Problem 10, for example, were based on the same observations, the probability of malignancy should be lower than in the case where the experts' estimates were based on independent observations. The empirical data indicate that the manner in which people integrate evidence depends on whether the data is causal or not, and is not very sensitive to the degree of dependency that determines their actual diagnostic value. More specifically, people appear to compound causal data and average data that is not causal, such as experts' estimates. This tendency leads to a major underestimation of the impact of non-causal evidence, which could have severe consequences in the intuitive assessment of legal, medical, or scientific evidence.

A demonstration of the averaging fallacy in the evaluation of scientific data was reported in Tversky and Kahneman (1971). A survey of statistically sophisticated research psychologists showed that experienced investigators are more impressed by a single experimental result which provides strong support for a hypothesis than by the conjunction of the same result with another, which provides positive but weaker support for that hypothesis. Evidently, the respondents averaged the degrees of support provided by the separate results instead of compounding them appropriately. The averaging fallacy, apparently, is not eradicated by training in statistics.

IV. DISCUSSION

According to the modern theory of subjective probability (DeFinetti, 1937; Savage, 1954), judgments of likelihood are viewed as personal expressions of degree of belief. If the judgments satisfy the axioms of the theory then they can be regarded as a coherent system of beliefs that is expressible by a probability measure. The axioms of the theory, however, do not provide the person with procedures to assess his degree of belief, or to translate his knowledge regarding an uncertain event into a number between zero and one. Moreover, the axiomatic theory is silent on the question of how to modify a set of inconsistent assessments so as to achieve coherence. Thus, the normative theory provides criteria for testing the consistency of judgments, but it provides no guidance on how to assess probabilities and resolve inconsistencies.

Recent psychological research on judgment under uncertainty has been concerned with the methods that people use to evaluate evidence, and the procedures on which they rely to assess subjective probability. We have suggested that people assess the probability of an event by using a limited number of basic computational procedures, or heuristics. Specifically, we have argued that in many contexts judgments of probability rely on an impression of similiarity or representativeness (Kahneman and Tversky, 1972, 1973), or on the availability and imaginability of instances (Tversky and Kahneman, 1972). In the present paper we have shown that judgments of probability are sometimes determined by impressions of causal impact. When a person relies on representativeness, availability or causality to assess subjective probability, he does not necessarily identify probability with any of these notions; he may simply use them as heuristics. These modes of judgment provide useful methods for assessing probability but they do not guarantee the consistency or coherence of the results, because they systematically depart from the laws of probability theory. Indeed, many of the biases and errors in judgments of probability can be traced to the incompatibility between the rules of probability and the rules that govern impressions of similarity, availability and causality.

People are often called upon not only to produce and interpret probability statements, but also to integrate or combine probabilistic statements that are presented to them in numerical or verbal form. The calculus of probability provides standard methods for combining such data, e.g. Bayes' rule. These computational methods take into account the informative value of each datum, and the dependencies among them. The source and nature of the data are immaterial. The application of Bayes' rule, for example, does not depend on whether the base-rate evidence does or does not have a causal interpretation. Intuitive judgments do not follow the correct rules of integration because the manner in which evidence is combined depends primarily on its nature, not only on its informative value.

Specifically, three rules of combination of probabilistic data have been illustrated in the present paper: dominance, compounding and averaging. The results of Section I showed that the causal component of conditional probability dominates the diagnostic component in problems where the two components should be averaged. A more direct demonstration of dominance was presented in Section II which showed that base-rate information is completely dominated by specific information, except when the base-rate is interpreted in causal terms. Finally, Section III indicated that probabilistic data are compounded when they are interpreted as propensities, and averaged when they are interpreted as estimates.

These observations illustrate the discrepancy between intuitive judgments under uncertainty and the normative theory of subjective probability. In the latter, all probability assessments refer to one's degree of belief, and the same formal calculus applies to the combination of all probabilistic data. In contrast, people have several different ways of thinking about probability, and their approach to the combination of probabilistic data depends on the manner in which those data are interpreted. This discrepancy presents a major educational challenge: to help people achieve a synthesis between their natural modes of judgment and the logic of probability.

Hebrew University, Jerusalem

NOTE

[1] Moreover, in Problems 9 and 10, conditional independence, i.e. $P(D_1$ and $D_2/H) = P(D_1/H)P(D_2/H)$, may be expected to hold, at least as a reasonable approximation. In this case, $P(H/D_1$ and $D_2)$ can be computed as follows. Let $\bar{H}$ denote the complement of H, hence

$$\frac{P(H/D_1 \text{ and } D_2)}{P(\bar{H}/D_1 \text{ and } D_2)} = \frac{P(D_1 \text{ and } D_2/H)P(H)}{P(D_1 \text{ and } D_2/\bar{H})P(\bar{H})} \quad \text{by Bayes' rule}$$

$$= \frac{P(D_1/H)P(D_2/H)P(H)}{P(D_1/\bar{H})P(D_2/\bar{H})P(\bar{H})} \quad \begin{array}{l}\text{by conditional} \\ \text{independence, for both } H \text{ and } \bar{H}\end{array}$$

$$= \frac{\dfrac{P(H/D_1)P(D_1)}{P(H)} \dfrac{P(H/D_2)P(D_2)}{P(H)} P(H)}{\dfrac{P(\bar{H}/D_1)P(D_1)}{P(\bar{H})} \dfrac{P(\bar{H}/D_2)P(D_2)}{P(\bar{H})} P(\bar{H})} \quad \text{by Bayes' rule}$$

$$= \frac{P(H/D_1)P(H/D_2)P(\bar{H})}{P(\bar{H}/D_1)P(\bar{H}/D_2)P(H)} .$$

Substituting the numerical values from Problems 9 and 10 yields

$$\frac{P(H/D_1 \text{ and } D_2)}{P(\bar{H}/D_1 \text{ and } D_2)} = \frac{(0.10)(0.45)(0.95)}{(0.90)(0.55)(0.05)} = \frac{19}{11}$$

and

$$P(H/D_1 \text{ and } D_2) = 0.63.$$

BIBLIOGRAPHY

Bar-Hillel, M., 'The Base-rate Fallacy in Subjective Judgments of Probability', Unpublished doctoral dissertation, Hebrew University, 1975.

Borgida, E. and Nisbett, R. E., 'Abstract vs Concrete Information: The Senses Engulf the Mind, Unpublished manuscript, 1976, University of Michigan.

Darley, J. M. and Latané, B., 'Bystander Intervention in Emergencies: Diffusion of Responsibility', *Journal of Personality and Social Psychology* **8** (1968), 377-383.

DeFinetti, B., 'La prévision: ses lois logiques, ses sources subjectives', *Ann. Inst. H. Poincaré* **7** (1937), 1-68. Translated into English in H. E. Kyburg, Jr. and H. E. Smokler (eds.) *Studies in Subjective Probability*, Wiley, New York, 1964, pp. 93-158.

Edwards, W., 'Conservatism in Human Information Processing', in B. Kleinmuntz (ed.), *Formal Representation of Human Judgment*, Wiley, New York, 1968, pp. 17-52.

Hammerton, M., 'A Case of Radical Probability Estimation', *Journal of Experimental Psychology* **101** (1973), 242-254.

Kahneman, D. and Tversky, A., 'Subjective Probability: A Judgment of Representativeness', *Cognitive Psychology* **3** (1972), 430-454.

Kahneman, D. and Tversky, A., 'On the Psychology of Prediction', *Psychological Review* **80** (1973), 237-251.

Lyon, D. and Slovic, P., 'Dominance of Accuracy Information and Neglect of Base Rates in Probability Estimation', *Acta Psychologica* **40** (1976), 287–298.

Nisbett, R. E. and Borgida, E., 'Attribution and the Psychology of Prediction', *Journal of Personality and Social Psychology* **32** (1975), 932–943.

Salmon, W. C., 'Confirmation and Relevance', in G. Maxwell and R. M. Anderson, Jr. (eds.), *Induction, Probability, and Confirmation*, University of Minnesota Press, Minneapolis 1975.

Savage, L. J., *The Foundations of Statistics*, Wiley, New York, 1954.

Slovic, P., 'From Shakespeare to Simon: Speculations- and Some Evidence- about Man's Ability to Process Information', *ORI Research Monograph* **12** (1972), 2, Oregon Research Institute.

Slovic, P., Fischhoff, B., and Lichtenstein, S., 'Cognitive Processes and Societal Risk Taking', in J. S. Carroll and J. W. Payne (eds.), *Cognition and Social Behavior*, Lawrence Erlbaum Associates, Potomac, Md., 1976.

Turoff, M., 'An Alternative Approach to Cross-impact Analysis', *Technological Forecasting and Social Change* **3** (1972), 309–339.

Tversky, A. and Kahneman, D., 'Belief in the Law of Small Numbers', *Psychological Bulletin* **76** (1971), 105-110.

Tversky, A. and Kahneman, D., 'Availability: A Heuristic for Judging Frequency and Probability', *Cognitive Psychology* **5** (1973), 207-232.

Tversky, A. and Kahneman, D., 'Judgment under Uncertainty: Heuristics and Biases', *Science* **185** (1974), 1124-1131.

IV

THE CONCEPT OF RANDOMNESS

C. P. SCHNORR

A SURVEY OF THE THEORY OF
RANDOM SEQUENCES

1. THE KOLMOGOROV CONCEPT OF FINITE
RANDOM SEQUENCES

There are now 10 years that the theory of random sequences revived beginning with a stimulating paper of Kolmogorov in 1965. In this paper Kolmogorov proposed a definition of finite random sequences with respect to the equiprobability distribution. Those binary sequences x were considered to be random for which the minimal length of a description for x (program complexity of x) differed little from the length of x. The idea that randomness is related to the minimal length of descriptions was independently developed in Chaitin (1966). By introducing the concept of a universal algorithm Kolmogorov gave a rigorous definition of program complexity which no more depends on the choice of a special machine (or algorithm).

Let X^* (X^∞, resp.) be the set of all finite (infinite resp.) binary sequences. Kolmogorov considered p.r. (partial recursive) functions $\psi: X^* \times N \to X^*$ such that

(1.1) $\qquad |\psi(x, n)| = n$ for all $(x, n) \in \text{domain}(\psi)$.

Hereby $|y|$ is the length of $y \in X^*$. If $\psi(y, |x|) = x$ then y is called a program (or description) of x relative to the length of x. Let

$$K_\psi(x) = \min\{|y|: \psi(y, |x|) = x\}$$

be the program complexity of x with respect to ψ. Hereby we set $\min \varnothing = \infty$.

A main observation of Kolmogorov was the existence of universal algorithms which yield minimal length descriptions (up to a bounded

Butts and Hintikka (eds.), Basic Problems in Methodology and Linguistics, 193–211.

additive term) for all sequences:

THEOREM (1.2) (Kolmogorov, 1965). *Among the p.r. functions that satisfy* (1.1) *there exists a universal* φ *such that for any* $\psi : \exists u \in X^*$: $\varphi(ux, n) = \psi(x, n)$ *for all* $(x, n) \in domain\ (\psi)$.

This implies

$$K_\varphi \lesssim K_\psi \quad \text{for all } \psi \text{ with (1.1)},$$

where $f \lesssim g$ means $f \le g + O(1)$.

In the following we set $K = K_\varphi$ for some fixed universal φ according to (1.2). $f \asymp g$ means $f \lesssim g$ and $g \lesssim f$.

Kolmogorov proposed to consider those sequences x as random with respect to the equiprobability distribution for which $|x| \asymp K(x)$, i.e. the difference of $|x|$ and $K(x)$ is bounded by some fixed constant. This means that the sequence x itself is almost a shortest description for x. The difference $|x| - K(x)$ seems to measure the randomness deficiency of x. Indeed those sequences x with $|x| \asymp K(x)$ have many random-like properties.

At first they constitute the antipode to the recursive sequences which can be characterised by their bounded program complexity. For $z \in X^\infty$ let z^n be the initial segment of length n. Then

PROPOSITION 1.3. $z \in X^\infty$ *is recursive iff* $K(z^n)$ *is bounded by some constant, i.e.* $\sup_n K(z^n) < \infty$.

Moreover, zero's and one's are equally distributed in these sequences. For $c \in N$ define

$$\gamma_c(n) = \max \left\{ \left| \sum_{i=1}^n x_i - \frac{n}{2} \right| : x \in X^n \wedge |K(x) - |x|| \le c \right\}$$

PROPOSITION 1.4. $\lim_n \gamma_c(n)/n = 0$ *for all* $c \in N$.

It is even possible to prove stronger assertions: γ_c has the limit behaviour which corresponds to the law of the iterated logarithm:

$$\lim_n \frac{2\gamma_c(n) - n}{\sqrt{2n \lg \lg n}} = 1$$

It turns out that almost all sequences of any fixed length are approximately random:

PROPOSITION 1.5 (Kolmogorov, 1965).

(a) $K(x) \lesssim |x|$

(b) $\|\{x \in X^n : K(x) < |x| - c\}\| \leq 2^{n-c}$

Hence the fraction of those sequences $x \in X^n$ for which the randomness deficiency $|x| - K(x)$ is greater than c is less than 2^{-c}.

Let us discuss some objections to this concept of randomness. It has been objected that a different choice of the universal algorithm φ changes the program complexity by some bounded additive term which however may be arbitrarily large. So the measure $K(x)$ might be meaningless for any specific finite sequence x. However, one can well argue that in all real computing machines, programs for the same problems have approximately the same size. Therefore we even can get rid of this bounded additive term if there is a canonical model for real computing machines. Such a model which has widely been accepted and which is superior to the classical Turing machine is the model of Random Access Machine (which only has the successor function instead of full addition and subtraction). This model is equivalent to the Kolmogorov-Uspensky machine which later on also has been proposed by Schönhage; see Schnorr (1974) for these equivalences. However, the original form of Kolmogorov's approach causes some more serious difficulties which lie in the behaviour of $K(x)$. Martin-Löf (1971) proved that the randomness deficiency oscillates in a strange way on the initial segments of almost all sequences:

THEOREM 1.6 (Martin-Löf, 1971).   *Let $f: N \to N$ be any rec. function such that $\sum_n 2^{-f(n)} = \infty$ (e.g. $f(n) = [\lg n]$). Then*

$$\forall z \in X^\infty : \exists \overset{\infty}{n} : n - K(z^n) \geq f(n).$$

In particular there is no infinite sequence $z \in X^\infty$ such that the randomness deficiency $|n - K(z^n)|$ is bounded. Therefore, this concept is not suitable for infinite sequences. 1.6 is most suitably explained by the following law concerning the equiprobability distribution μ on infinite binary sequences which can be proved by using the Borel Cantelli-Lemma (Let z_i be the i-th element of $z = z_1 z_2 \cdots \in X^\infty$):

THEOREM 1.7.   *Let* $f: N \to N$ *be any rec. function such that* $\sum_n 2^{-f(n)} = \infty$. *Then* $\mu\{z \in X^\infty \mid \overset{\infty}{\exists} n : \forall i \leq f(n) : z_{n-i} = 0\} = 1$

Observe that $\forall i \leq f(n): z_{n-i} = 0$ implies $K(z^n) \overset{\textstyle <}{\sim} n - f(n)$. For a sequence $x \in X^n$ with a series of $f(n)$ zero's at the end can be described by combining a program for $x^{n-f(n)}$ and a program for f. However, because of (1.7) a series of $f(n)$ zero's in a sequence $x \in X^n$ does not generally indicate a defect of randomness. This shows that the measure $K(x)$ exploits the position of a series of zero's relative to the right end of x. Since a finite random sequence is always an initial segment of some infinite random sequence we conclude that the position of a series of zero's relative to the right end of sequence x cannot be related at all to the random behaviour of x. This shows that $|x| \overset{\textstyle <}{\times} K(x)$ is too strong a condition for randomness. Therefore we propose to call those sequences x with $|x| \overset{\textstyle <}{\times} K(x)$ *patternless* sequences in order to distinguish them from random sequences.

In order to overcome these difficulties we shall modify the original approach of program complexity. We require that an appropriate measure for the randomness deficiency preserves regularities of initial segments. This is not true for $|x| - K(x)$ since $|xy| - K(xy)$ may be small even if $|x| - K(x)$ is a large number. The concept of monotonic complexity Km in Section 4 levels the oscillations of $|x| - K(x)$ and it preserves regularities of initial segments since $|x| - Km(x)$ is almost non-decreasing in the following sense (see also 3.9):

$$|xy| - Km(xy) \overset{\textstyle >}{\sim} |x| - Km(x) - (1 + \varepsilon) lg(|x| - Km(x))$$

for all $\varepsilon > o$.

We should note that all variants of program complexity that are considered in this paper such as K, Km, lgU_μ, KP only slightly differ one from the other. They coincide to within the logarithm of the minimum of all these measures. 1.7 implies that none of these measures is invariant with respect to cyclic shifts of sequences and this invariance cannot be expected from any complexity measure at all.

2. THE DEFINITIONS OF INFINITE RANDOM SEQUENCES
BY MARTIN-LÖF AND SCHNORR

In order to avoid the difficulties with Kolmogorov's approach Martin-Löf (1966) proposed a definition of infinite random sequences ('Kollektive') which is based on constructive measure theory. The basic idea is that in probability theory random properties of infinite sequences are expressed by sets of measure zero (null-sets). Therefore infinite random sequences can be defined by formalizing the class of those null-sets that correspond to random properties. Of course these null-sets should be constructive in some sense.

Let p be a *c.p.m.* (computable probability measure) on X^∞, i.e. p is given by a computable (in the sense of recursive analysis) function $p: X^* \to [0, 1]$ such that

(1) $\forall x \in X^*: p(x0) + p(x1) = p(x)$

(2) $p(\Lambda) = 1$ for the empty sequence Λ.

p can be extended in a canonical way to a σ-additive measure on X^∞ which we also denote p.

DEFINITION 2.1 (Martin-Löf, 1966). A set $\mathfrak{N} \subset X^\infty$ is a *recursive p-null-set* iff there is a r.e. (rec. enumerable) set $Y \subset N \times X^*$ such that for $Y_i := \{x | (x, i) \in Y\}$:

(1) $\forall i: p[Y_i] \le 2^{-i}$ (2) $\mathfrak{N} \subset \bigcap_{i \in N} [Y_i]$

Hereby $[A] = AX^\infty$ is the cylinder that is associated with $A \subset X^*$.

A sequence $z \in X^\infty$ is *Martin-Löf p-random* (M.L. p-random) iff z is not contained in any rec. p-null-set. Let $\mathcal{K}^p_{M.L.}$ be the set of all M.L. p-random sequences.

Inspection of those examples of null-sets that are known from probability theory such as the strong law of large numbers and the law of the iterated logarithm show that all these examples satisfy the condition that critical regions of measure $\le 2^{-i}$ can be recursively enumerated. So we can expect that all null-sets that express random properties are recursive in the sense of 2.1. However, we have to be careful on the question whether any sequence z which is contained in some rec. null set $\mathfrak{N}$ cannot be accepted as a random sequence. It is not clear what $z \in \bigcap_i [Y_i]$ with Y as in 2.1 means for the behaviour of the initial segments of z since we cannot effectively compute $p([Y_i] \cap [z^n])$. We argue that the rec. null-sets might be a too general class of null-sets. Our object is not to find the most general class of null-sets that can be constructively described in some way. Note that the class of rec. null-sets can easily be generalized by using higher order arithmetical (or analytical predicates, see e.g. Martin-Löf, 1970). But the problem is to formalize exactly the class of those null-sets that are 'effective' in the same sense that the recursive functions are 'the effective functions' by Church's thesis. We argue that this is a proper subclass of the rec. null-sets and in the following we propose a definition for this class and we like to give evidence for this concept.

One main feature of the class of rec. null-sets that contrasts to the following concept of total recursive null-sets is the existence of a universal null-set which contains all others:

THEOREM 2.2 (Martin-Löf, 1966).   *For any c.p.m. p. the union of all rec. p-null-sets $X^\infty - \mathcal{H}^p_{M.L.}$ is a (universal) rec. p-null-set.*

This also implies $p(\mathcal{H}^p_{M.L.}) = 1$ which corresponds to 1.5(b) and means that almost all sequences are random.

In contrast to the above concept Schnorr (1969, 1971) proposed a stronger concept of constructive null-set: $\mathfrak{N}$ is called a *total recursive p-null-set* if there is an r.e. set Y such that in addition to 2.1 the function f with $f(i) = p[Y_i]$ is a computable function. A sequence is called *Schnorr p-random* iff it is not contained in any total recursive p-null-set. Let $\mathcal{H}^p_{Sch}$ be the set of all Schnorr p-random sequences.

Indeed an inspection of the proofs for the standard laws of large numbers in probability theory shows that these proofs implicitly yield algorithms which compute the measure $p[Y_i]$ of critical regions for these null-sets. So even this restricted class of null-sets seems to be general enough to express all random properties. On the other hand we can prove theorems which strongly suggest that any sequence which is not Schnorr p-random does not behave like a p-random sequence, see for instance 3.2. Also note that our notion of total rec. null-sets is the recursive analogue to the intuitionistic notion of sets of measure zero by Brouwer, see Heyting (1956). This yields the following

THESIS 2.3.   A sequence behaves within all effective procedures like a p-random sequence iff it is Schnorr p-random.

In general the class $\mathcal{K}^p_{M.L.}$ is properly contained in the class $\mathcal{K}^p_{Sch}$ of Schnorr p-random sequences (2.4). This fact seems to be important for the problem of approximating random behaviour by effective methods. It is easier to approximate sequences in the larger set $\mathcal{K}^p_{Sch}$, see for instance 3.6.

$z \in X^\infty$ is called an *atom* with respect to p if $p(z) > 0$. p is called *discrete* if the set of atoms of p has p-measure 1. Obviously all atoms of every *c.p.m.* p are recursive.

THEOREM 2.4.   (1) $\mathcal{K}^p_{M.L.} \subset \mathcal{K}^p_{Sch}$ *for all c.p.m.* p
(2) $\mathcal{K}^p_{M.L.} = \mathcal{K}^p_{Sch}$ *iff* p *is discrete.*

To prove the 'only if-part' of (2) one has to generalize a construction in Schnorr (1969) which handles the case of the equiprobability distribution.

The following theorem also contrasts to the Martin-Löf concept.

THEOREM 2.5 (Schnorr, 1971).   *Let* $\mathfrak{N}$ *be any total rec. p-null set, then there exists a rec. sequence* z *such that* $z \notin \mathfrak{N}$.

In the standard case where p has no atoms there does not exist a rec. p-random sequence:

PROPOSITION 2.6.   *Every rec. sequence in* $\mathcal{K}^p_{Sch}$ *is an atom of p.*

If p is not discrete than there exists no universal total rec. null-set which contains all others. However, the union of an r.e. set of total rec. null-sets is itself a total rec. null-set, see Schnorr (1971).

We think that it is an interesting observation that rec. sequences are a special case of random sequences:

PROPOSITION 2.7.   *For every rec. sequence* z *there is a rec. prob. measure* $p: X^* \rightarrow \{0, 1\}$ *such that* $\{z\} = \mathcal{K}^p_{Sch} = \mathcal{K}^p_{M.L.}$

This suggests yields to the following definition of general random sequences:

DEFINITION 2.8 (Schnorr et al., 1975).   The set $\mathcal{K}_{Sch}$ ($\mathcal{K}_{M.L.}$, resp.) of all *general* Schnorr (Martin-Löf, resp.) random sequences is the union of all sets $\mathcal{K}^p_{Sch}$ ($\mathcal{K}^p_{M.L.}$, resp.) where p ranges over all *c.p.m.*'s.

There are many connections between these concepts of infinite random sequences and the Kolmogorov approach, see e.g. Schnorr

(1971). Fuchs (1975) has probably given the most natural description of $\mathcal{K}^{\mu}_{Sch}$ and $\mathcal{K}^{\mu}_{M.L.}$ (where μ is the equiprobability distribution) in terms of the Kolmogorov complexity K.

One advantage of the above approach to randomness of infinite sequences is that it easily points out connections to many strong laws of large numbers in probability theory. However, we think that this approach does not yield the most intuitive definition of randomness and certainly not the most general one. For instance, there is no direct way to extend this concept to non-computable distributions p. However, we do not know the algorithmical structure of the distributions that occur in nature. Therefore, there should be an approach to randomness which does not require any computability assumptions on p and moreover the set of p-random sequences should not depend on the algorithmical structure of p. Such an approach is given in the next section.

3. RANDOM SEQUENCES AND THE PRINCIPLE OF THE EXCLUSION OF GAMBLING SYSTEMS

In 1919 von Mises proposed a concept of 'Kollektive' using the principle that there should be no gambling system that permits unrestricted profits when gambling on these sequences. v. Mises required that the relative frequency of 0's and 1's in a Kollektiv z should be the same as in any subsequence of z that can be obtained from z by a selection rule. Church (1940) gave a precise definition of this concept based on recursive selection rules. However, Ville (1939) proved that the concept of selection rules is to weak for a suitable concept of random sequences. He showed that any countable system of selection rules admits random sequences z such that each segment z^n has more 0's than 1's. We shall use a more general type of gambling systems than selection rules. Martingales are functions that describe the capital of a gambler when playing on binary sequences. One of the main results in Schnorr (1971) states that there is a natural characterization of random sequences by martingales. The profit that a gambler can make when playing on the sequence $z \in X^{\infty}$ is a measure for the randomness deficiency of z.

Let p be any (not necessarily computable) probability measure on X^{∞}. Then a c. (computable) p-martingale is a function $V: X^* \to R \cup \{\infty\}$

such that the product $p \cdot V: X^* \rightarrow [0, 1]$ is a *c.p.m.* The martingale property

(3.1) $p(x)V(x) = p(x0)V(x0) + p(x1)V(x1)$

is the condition of a fair play and means that a player has the same chance to win as to lose when he plays on sequences that are distributed according to p. $V(x)$ is the total capital of the gambler after playing on the sequence $x \in X^*$. He always starts with the unit capital $V(\Lambda) = 1$ at the beginning of the game. If p is computable then instead of requiring that $p \cdot V$ is computable we could as well require that V is computable. This, however, makes no sense if p is not computable since any computable V with (3.1) would be constant in this case. Our martingale approach to randomness will not use any computability assumptions on p. In the following we shall always assume that a *c.p*-martingale V is given by a program for $p \cdot V$. (One can use the convention $V(x) = \infty$ whenever $p(x) = 0$ in order to determine V from $p \cdot V$ and p.)

Let $\mathscr{L}$ be the class of all recursive, non-decreasing and unbounded functions $f: N \rightarrow N$. The following characterizations of $\mathscr{K}^p_{Sch}$ and $\mathscr{K}^p_{M.L.}$ hold for all *c.p.m.*'s, see Schnorr (1971):

THEOREM 3.2. $z \notin \mathscr{K}^p_{Sch}$ *iff there is a c.p-martingale V and $g \in \mathscr{L}$ such that* $\overline{\lim\limits_n} V(z^n)/g(n) > 0$.

This means z is not Schnorr p-random iff there is a *c.p.*-martingale V which is 'constructively' unbounded on z.

THEOREM 3.3. $z \notin \mathscr{K}^p_{M.L.}$ *iff there is a semi-computable p-submartingale V such that* $\overline{\lim\limits_n} V(z^n) = \infty$.

Hereby a *semi-computable p-submartingale* is a function V such that $p(x)V(x) \geq p(x0)V(x0) + p(x1)V(x1)$ for which there is a recursive rational function

$$g: N \times X^* \rightarrow Q \text{ such that } p(x)V(x) = \sup_n g(n, x) \text{ for all } x \in X^*.$$

Theorems 3.2 and 3.3 clearly point out the difference between the Martin-Löf and the Schnorr concept of infinite random sequences.

There are non-effective gambling systems which are involved in the Martin-Löf approach.

Because of theorems 3.2 and 3.3 we can extend the definition of random sequences to non-computable prob. measures:

DEFINITION 3.4. For any non-comp. prob. measure p define the classes $\mathcal{K}^p_{Sch}$ ($\mathcal{K}^p_{M.L.}$, resp.) such that the assertion of theorem 3.2 (3.3, resp.) holds.

This definition is also motivated by the following observation: V is a $c.p$-martingale iff $V\bar{p}/p$ is a $c.\bar{p}$-martingale. For any $c.p$-martingale V and $z \in X^\infty$ the function f with $f(n) = \lg V(z^n)$ measures the p-randomness deficiency of z and is called a p-defect of z. Let $DEF_p(z)$ be the set of all p-defects of z with respect to all $c.p$-martingales.

Obviously the set $DEF_p(z)$ describes the p-random behaviour of z. But $DEF_p(z)$ also describes the p-random behaviour of z for any p.m. $\bar{p}$. For the set $DEF_p(z)$ of functions differs from $DEF_{\bar{p}}(z)$ only by the additive term $\lg \bar{p}(z^n) - \lg p(z^n)$. Another immediate consequence of this observation is a characterization of the general Schnorr random sequences as those sequences for which there exists an optimal gambling system in terms of martingales (3.5).

The following relation $\alpha \leq^* \beta$ for functions $\alpha, \beta : N \to R$ means that $\beta - \alpha$ is not 'constructively' unbounded:

$$\alpha \leq^* \beta \Leftrightarrow \forall g \in \mathcal{L} : \alpha \stackrel{\sim}{\leq} \beta + g.$$

PROPOSITION 3.5. *The following assertions are equivalent:*
 (1) $z \in \mathcal{K}_{Sch}$ ($\mathcal{K}_{M.L.}$, resp.)
 (2) z *admits a* $\leq^*$-*maximal* ($\stackrel{\sim}{\leq}$-*maximal, resp.*) p-*defect for all prob. measures* p, *i.e.*:

$$\exists \alpha \in DEF_p(z) : \forall \beta \in DEF_p(z) : \beta \leq^* \alpha \quad (\beta \stackrel{\sim}{\leq} \alpha, \text{ resp.})$$

It should be interesting to study the possible structure of the rapidly growing functions in $DEF_p(z)$. Another interesting point is the dependence of the functions in $DEF_p(z)$ on the resource bounds of the programs for $p \cdot V$.

Let p be a c.p.m. Then it can easily be seen that we may restrict to those $c.p$-martingales V such that $V : X^* \to Q$ is a recursive, rational function. We say that V has time bound T with $T \in \mathcal{L}$ if there is a program that computes $V(x)$ within $T(|x|)$ steps. The following theorem states that p-pseudo-random sequences can effectively be constructed:

THEOREM 3.6 (Schnorr, 1971). *For any c.p.m. p and any $T \in \mathcal{L}$ there exists a recursive sequence z such that $\sup_n V(z^n) < \infty$ for all p-martingales V with time bound T.*

There is also an interesting connexion to Church random sequences which shows that Church random sequences approximate the behaviour of Schnorr random sequences. Remember that a sequence $z \in X^\infty$ is Church-random with respect to the equiprobability measure μ if $\lim_n \sum_{i=1}^n z_i / n = \frac{1}{2}$ and if this limit-property holds for all infinite subsequences which can be obtained from z by recursive selection rules. Our main result is the following

THEOREM 3.7 (Schnorr, 1971). *Let z be Church μ-random then $\lim_n \beta(n)/n = 0$ for all $\beta \in DEF_\mu(z)$ i.e. there is no μ-martingale which increases exponentially on z.*

Theorems 3.6 and 3.7 are related to two different types of approximations of random sequences and there are two different corresponding classifications of the total rec. null-sets.

Theorem 3.6 refers to a classification of the $c.p$-martingales by their time bound which yields an infinite hierarchy of total rec. null-sets. On the other hand we can classify the null-sets by the growth-rate of martingales.

A p-null-set $\mathfrak{N}$ is called to be of order $T \in \mathcal{L}$ if there is a p-martingale V such that

$$\mathfrak{N} \subset \{z \mid \varlimsup_n V(z^n)/T(n) > 0\}.$$

The important fact behind theorem 3.7 is that all μ-null-sets of exponential order can be obtained by trivial transformations (using selection rules) from the strong law of large numbers which itself is of 'almost' exponential order. It is an open problem whether some stronger laws such as for instance the law of the iterated logarithm will generate in the same way all μ-null-sets of linear order. These classifications of the laws of probability seem to reflect the importance

of statistical laws, for instance the strong law of large numbers is of
high order and low complexity. The classification of null-sets according
to the growth rate of martingales also yields an infinite hierarchy, for
details on the above hierarchies see Schnorr (1971).

Let us now consider some consequences of the martingale approach
to the Martin-Löf concept of randomness. Using standard techniques
we can recursively enumerate all semi-computable p-submartingales
provided p is computable. This yields the existence of a universal
semi-comp. p-submartingale:

PROPOSITION 3.8. *Let p be any c.p.m. p, then there exists a universal semi-comp. p-submartingale U_p such that for any semi-comp. p-
submartingale V:* $\lg V \stackrel{\scriptstyle <}{\scriptstyle \sim} \lg U_p$.

This implies $z \in \mathcal{K}_{M.L.}^p$ iff $\overline{\lim_n} U_p(z^n) < \infty$. Observe that we can choose

U_p such that $pU_p = \bar{p}U_{\bar{p}}$ for all $p, \bar{p}$. pU_p coincides with the *a priori*
probability R which Levin introduced in Zvonkin *et al.* (1970).

The universal p-submartingale U_p yields a *universal p-defect*
$\lg U_p(z^n)$ for all sequences $z \in X^\infty$. This universal p-defect is 'almost'
non-decreasing in the following sense:

THEOREM 3.9

$$\lg U_p(xy) \stackrel{\scriptstyle >}{\scriptstyle \sim} \lg U_p(x) - (1+\varepsilon) \lg \lg U_p(x) \quad \text{for all} \quad \varepsilon > 0.$$

$\lg U_p(x)$ is non-decreasing up to the low order term $(1 + \varepsilon) \lg \lg U_p(x)$. Therefore, in this measure of complexity regularities of
initial segments of sequences are preserved: if x has regularities with
respect to the c.p.m. p then all sequences in xX^* have regularities, too.
This explains why the universal p-deficiency $\lg U_p$ does not oscillate as
it is the case with $|x| - K(x)$, see (1.6).

Proof of 3.9. With U_p we can associate the following semi-
computable p-submartingales U_p^i and $\tilde{U}_p$:

$$U_p^i(x) = \begin{cases} 2^i & \text{if } \exists j \leqslant |x|: U_p(x^j) > 2^i \\ 2^i p\{xz \in X^\infty | \sup U_p(xz^n) > 2^i\} & \text{otherwise} \end{cases}$$

$$\tilde{U}_p = \sum_i i^{-1-\varepsilon} U_p^i$$

Observe that $U_p(\Lambda) \leqslant 1$ implies $U_p^i(\Lambda) \leqslant 1$ and $\tilde{U}_p(\Lambda) < \infty$. It follows:

$$\lg U_p(xy) \gtrsim \lg \tilde{U}_p(xy) \geqslant \lg U_p(x) - (1 + \varepsilon) \lg i$$

for all i with $2^i < U_p(x)$ ∎

4. MONOTONIC PROGRAM COMPLEXITY AND THE CONCEPT OF LEARNABLE SEQUENCES

The difficulties with the original Kolmogorov concept of program complexity can be eliminated by imposing a monotonicity condition on the algorithms that are used to describe binary sequences. Such conditions have independently been developed by Levin (1973) and Schnorr (1973).

A *p.r.* (partial recursive) function $\varphi: X^* \times N \to X^*$ is called *monotonic* iff

(1) $\forall (x, n) \in domain\ (\varphi): |\varphi(x, n)| = n$

(2) $\forall (x, n), (xy, n + k) \in domain\ (\varphi) \cdot \varphi(x, n) \subset \varphi(xy, n + k)$

Here $u \subset v$ means that u is an initial segment of v.

Then $Km_\varphi(x) = \min \{|y|: \varphi(y, |x|) = x\}$ is the (unrestricted) monotonic complexity of $x \in X^*$ with respect to φ.

PROPOSITION 4.1. *There exists a universal p.r. monotonic φ such that for every p.r. monotonic ψ there exists $u \in x^*$ such that $\psi(x, n) = \varphi(ux, n)$ for all $x \in X^*$.*

Let φ be a fixed universal *p.r.* monotonic φ according to Proposition 4.1. We set $Km = Km_\varphi$. Obviously $Km \lesssim Km_\psi$ for all *p.r.* monotonic ψ.

THEOREM 4.2 (Levin, 1973; Schnorr, 1973). *For any prob. measure $p: z \in \mathcal{K}_{M.L.}^p$ iff $Km(z^n) \lesssim -\lg p(z^n)$.*

The monotonic program complexity not only yields a suitable characterization of random sequences. The following concept of monotonic program complexity with restricted resource bound describes the random behaviour of all sequences z in the same way as the sets $DEF_p(z)$ of all p-defects of z.

Let φ be a universal function as in Proposition 4.1 and let ϕ be a running time for φ in the sense of Blum (1967), i.e. *domain* $(\varphi) =$

domain (ϕ) and the predicate $\lambda x, n, m. \phi(x,n) = m$ is recursive. Then we define the monotonic program complexity with resource bound T:

$$Km^T(x) = \min \left\{ |y| : \begin{matrix} \forall i \leq |x| : \exists k \leq |y| : \\ \varphi(y^k, i) = x^i \wedge \phi(y^k, i) \leq T(i) \end{matrix} \right\}.$$

$-\lg p(x) - Km^T(x)$ measures the p-randomness deficiency of x in the same way as $\lg V(x)$ for c.p-martingales V:

THEOREM 4.3 (Schnorr *et al.*, 1975).

(1) *for every* $T \in \mathcal{L}$ *there is a c.p-martingale* V *such that*

$$\lg V(x) \lessapprox -\lg p(x) - Km^T(x) \quad \text{for all} \quad x \in X^*.$$

(2) *for every c.p-martingale* V *there exists* $T \in \mathcal{L}$ *such that*

$$\lg V(x) \lessapprox -\lg p(x) - Km^T(x).$$

Obviously we can translate Theorem 3.2 by using Theorem 4.3 in order to characterize Schnorr p-random sequences in terms of the monotonic program complexity.

The function $-\lg p(x) - Km(x)$ describes the p-randomness deficiency in a similar way as the universal p-defect $\lg U_p(x)$. $-\lg p(x) - Km(x)$ is 'almost' non-decreasing in the sense of Theorem 3.9, see Schnorr (1973).

It is not difficult to see that the randomness deficiencies $-\lg p(x) - Km(x)$, $\lg U_p(x)$ coincide to within a term $\lg \lg U_p(x)$:

$$\lg U_p(x) \lessapprox -\lg p(x) - Km(x) \lessapprox \lg U_p(x) + 2 \lg \lg U_p(x).$$

However, Gacs (personal communication) proved that they do not coincide to within a constant additive term. The discrete scale that is associated with Km makes the measure $-\lg p(x) - Km(x)$ less flexible than $\lg U_p(x)$. This indicates that the universal randomness deficiency $\lg U_p(x)$ is a more fundamental measure than $-\lg p(x) - Km(x)$.

We shall now give a characterization of the general random sequences in terms of the restricted monotonic program complexity:

THEOREM 4.4 (Schnorr et al., 1975).

(1) $\quad z \in \mathcal{K}_{Sch} \Leftrightarrow \exists T \in \mathcal{L} : \forall \bar{T} \in \mathcal{L} : KM^T(z^n) \leq^* Km^{\bar{T}}(z^n)$

(2) $\quad z \in \mathcal{K}_{M.L.} \Leftrightarrow \exists T \in \mathcal{L} : Km^T(z^n) \lessapprox Km(z^n)$

The above characterizations express one interesting feature of general random sequences. The general random sequences are exactly those sequences z such that the entire algorithmical structure of z can be expressed by a single program. This program may compute a time bound $T \in \mathcal{L}$ such that the restricted complexity Km^T is minimal on z, or it may compute a probability measure p such that z is p-random. In a similar context Daley (1975) proposed to call those sequences *learnable* sequences.

5. REAL CONSTRUCTIVE PROBABILITY SPACES

If we consider real sequences instead of binary sequences then some additional problems arise. Our first task is to characterize the class of *c.p.m.'s* on $R(R^\infty$, resp.). The requirement that a real *c.p.m.* p yields a computable function $\lambda a, b \,.\, p[a, b]$ with respect to the endpoints a, b of intervals is too restrictive. Since all computable functions are continuous this requirement implies $p(r) = 0$ for all $r \in R$. A more general class of *c.p.m.'s* is obtained by requiring that p is a computable function on a suitable algebra of subsets of R. In the case of X^∞ such an algebra is canonically determined as the class of cylinders

$$B = \{[A] \mid A \subset X^* \quad \text{and} \quad A \text{ is finite}\}.$$

We now define the analogue with respect to real probability spaces. In this case the role of the cylinders $[x]$ with $x \in X^*$ is played by the computable intervals (i.e. intervals with computable endpoints which are open, closed, half-open, or single points). Let I_c be the algebra which is generated by all computable intervals.

DEFINITION 5.1. A (computable) basis of R is a computable function $\gamma: B \to I_c$ such that
  (1) γ is an algebra homomorphism with respect to union, intersection and complementation of sets.
  (2) there is a recursive function $\partial: B \to B$ (boundary) such that $\gamma\partial[A]$ is the topological boundary of $\gamma[A]$.
  (3) $\gamma[xA]$ is a computable interval for all $x \in X^*$.
  (4) let l be the length of intervals then $\lim_n l\gamma(z^n) \to 0$ for all $z \in X^\infty$.

DEFINITION 5.2. A σ-additive real probability measure p is called *computable* iff there exists a basis γ of R such that $p(\gamma): B \to R$ is a computable function.

For any real *c.p.m.* p with basis γ one defines in a standard way the class $\mathfrak{M}_p$ of all recursively p-measurable subsets of R. $\mathfrak{M}_p$ is the analogue to Brouwers notion of constructively measurable sets within the framework of intuitionism.

The properties of our concept of basis γ imply the following

PROPOSITION 5.3 (Meinhardt and Schnorr, 1974). *The class $\mathfrak{M}_p$ of recursively p-measurable sets does not depend on the choice of a basis for p.*

Let $\tilde{p}$ be the product measure on R^∞ of the real *c.p.m.* p. A basis γ for p yields in a canonical way a basis $\tilde{\gamma}$ for $\tilde{p}$. $\tilde{\gamma}$ takes elements of the algebra which is generated by the finite products of computable intervals.

DEFINITION 5.4. A sequence $z \in R^\infty$ is called p-random if it is not contained in any $\tilde{p}$-null-set.

The main result now shows that all sets A in $\mathfrak{M}_p$ have frequency $p(A)$ with respect to all $\tilde{p}$-random sequences.

THEOREM 5.5 (Meinhardt and Schnorr, 1974). *Let $A \in \mathfrak{M}_p$ and let z be $\tilde{p}$-random then*

$$\lim_n \frac{1}{n} \sum_{i=1}^{n} \chi_A(z_i) = p(A).$$

It is in this way that the concept of random sequences reconstructs a starting point of classical axiomatic probability theory. All constructively measurable sets have a probability which equals a relative frequency.

The concept of p-random sequences according to 5.4 is extremely strong in the following respect. It implies that a p-random sequence $r \in R^\infty$ has only components r_i that are not contained in any p-null-set. Hence all components of r are 'p-random'. In particular no component of r is computable when p is a continuous measure.

Therefore, Bos (1974) proposed a weaker concept of randomness which uses additional continuity requirements for p-random tests. He can prove that there always exist random sequences in his concept with all components being rational. However, the class of subsets of R which have a relative frequency with respect to all random sequences of the Bos-concept is decisively smaller. It coincides with the class of Jordan-Peano-measurable sets, i.e. all measurable sets A such that the topological boundary of A has measure 0.

6. CONNECTIONS WITH INFORMATION THEORY

The program complexity can be used to define a concept of algorithmical information of single objects. It can easily be seen that the information

$$I(x:y) = K(y) - K(y/x)$$

concerning the sequence y in the sequence x is symmetric to within a term $O(\lg I(x:y))$ but it is not symmetric to within a bounded term, for a proof see Zvonkin et al. (1970).

Recently another modification of the program complexity has independently been proposed by Levin (1974) and Chaitin (1975) which reflects information theoretical equalities to within a bounded additive term, see also the paper of Gacs (1974).

Let Y, Z be r.e. sets. A p.r. function $\psi: X^* \times Y \to Z$ is called *self-deliminating* (*s.d.*) if

(6.1) $\forall (x, y), (\bar{x}, y) \in domain\ (\psi): x \subset \bar{x} \Rightarrow x = x.$

If $\psi(x, y) = z$ then x is called a program for z relative to y. The property 6.1 means that if an infinite sequence $z \in X^\infty$ enters the machine M for ψ, then M must itself decide how many symbols of z are required for its computation.

There is a universal *s.d.* $B: X^* \times Y \to Z$ such that for any *s.d.* $\psi: X^* \times Y \to Z$ there exists $u \in X^*$:

$$B(ux, y) = \psi(x, y) \quad \text{for all} \quad (x, y) \in X^* \times Y.$$

Then

$$KP(z/y) := \min\{|x|: B(x, y) = z\}$$

is the program complexity of z relative to y and

$$KP(z) := KP(z/\Lambda)$$

is the complexity of z. Then the following analogue of the well-known equality

$$H(X, Y) = H(X) + H(Y \mid X)$$

concerning the Shannon entropy H of random variables X and Y holds (hereby $f \times g$ means $f \lesssim g$ and $g \lesssim f$):

THEOREM 6.2 (Levin, 1974; Chaitin, 1975).

$$KP(x, y) \times KP(x) + KP(y/(x, KP(x))).$$

Therefore, all information theoretical equalities hold for KP to within a bounded term if we always replace objects x in terms (y/x) by $(x, KP(x))$. Gacs (1974) proved that the information

$$IP(x:y) := KP(y) - KP(y/x)$$

is not itself asymptotically symmetric but it follows from 6.2 that

$$KP(y) - KP(y/(x, KP(x))) \times KP(x) - KP(x/(y, KP(y))).$$

University of Frankfurt, Fachbereich Mathematik,
6 Frankfurt a.M.

BIBLIOGRAPHY

Blum, M.: 1967, 'A Machnie-Independent Theory of Recursive Functions', *Journal ACM* **14.**
Bos, U.: 1974, 'Reelle Zufallsfolgen', Dissertation Frankfurt.
Chaitin, G. J.: 1966, 'On the Length of Programs for Computing Binary Sequences', *Journal ACM* **13.**

Chaitin, G. J.: 1975, 'A Theory of Program Size Formally Identical to Information Theory', *Journal ACM* **22.**

Church, A.: 1940, 'On the Concept of a Random Sequence', *Bull AMS* **46,** 130–135.

Daley, R.: 1975, 'On the Inference of Optimal Descriptions', Preprint University Chicago.

Fuchs, P.: 1975, 'Zufälligkeit, Programmkomplexität und Lernbarkeit', Dissertation Universität Frankfurt.

Gacs, P.: 1974, 'On the Symmetry of Algorithmic Information' *Soviet Math. Dokl.* **15.**

Heyting, A.: 1956, '*Intuitionism, an Introduction*' North-Holland Publ. Comp., Amsterdam, 1956.

Kolmogorov, A. N.: 1965, 'Three Approaches to the Quantitative Definition of Information', *Problems of Information Transmission* **1,** 1–7.

Levin, L. A.: 1973, 'On the Notion of a Random Sequence', *Soviet Math. Dokl.* **14.**

Levin, L. A.: 1974, 'Laws of Information Conservation Non-increase and Problems of the Foundation of Probability Theory', *Problemi Peredači Informacii* **10.**

Martin-Löf, P.: 1966, 'The Definition of Random Sequences', *Inform. Control,* **19,** 602–619.

Martin-Löf, P.: 1970, 'On the Notion of Randomness', in: *Intuitionism Proof Theory,* Proc. Summer Conf. Buffalo, N.Y. 1968, 73–78.

Martin-Löf, P.: 1971, 'Complexity Oscillations in Infinite Binary Sequences'. *Zeit. Wahrsch. verw. Geb.* **19,** 225–230.

Meinhardt, G.: 1974, 'Zufallsfolgen in reellen Wahrscheinlichkeitsräumen', Diplomarbeit Erlangen/Frankfurt.

Schnorr, C. P.: 1969, 'Eine Bemerkung zum Begriff der zufälligen Folge', *Zeit. Wahrsch. verw. Geb.* **14.**

Schnorr, C. P.: 1971, '*Zufälligkeit und Wahrscheinlichkeit*', Lecture Notes in Mathematics Vol. 218, Springer-Verlag, Berlin-New York, 1971.

Schnorr, C. P.: 1973, 'Process Complexity and Effective Random Tests', *J. Comp. System Sciences* **7.**

Schnorr, C. P.: 1974, *Rekursive Funktionen und ihre Komplexität*, Teubner Verlag, Stuttgart, 1974.

Schnorr, C. P., Fuchs, P.: 1975, 'General Random Sequences and the Concept of Learnable Sequences', Preprint Universität Frankfurt, to appear in *J. Symb. Logic.*

Ville, J.: 1939, *Étude critique de la notion de collectif*, Gauthiers-Villars, Paris, 1939.

Von Mises, R.: 1919, 'Grundlagen der Wahrscheinlichkeitstheorie', *Math. Zeit.* **5.**

Zvonkin, A. and Levin, L.: 1970, 'The Complexity of Finite Objects and the Development of the Concepts of Information and Randomness by Means of the Theory of Algorithms', *Russian Math. Surveys* **156,** 83–124.

RICHARD C. JEFFREY

MISES REDUX

Once one has clarified the concept of random sequence, one can define the probability of an event as the limit of the relative frequency with which this event occurs in the random sequence. This concept of probability then has a well defined physical interpretation. (Schnorr, 1971, pp. 8–9)

Mises' (1919) concept of *irregular* ('random') *sequence* resisted precise mathematical definition for over four decades. (See Martin-Löf, 1970, for some details.) This circumstance led many to see the difficulty of defining 'irregular' as *the* obstacle to success of Mises' program, and to suppose that the solution of that difficulty in recent years has finally set probability theory on the sure path of a science along lines that Mises had envisaged. To the contrary, I shall argue that since stochastic processes do not go on forever, Mises' identification of each such processes with *the infinite sequence of outputs it would produce if it ran forever* is a metaphysical conceit that provides no physical interpretation of probability.

1. BERNOULLI TRIALS

Martin-Löf (1966) showed how to overcome the distracting technical obstacle to Mises' program, and Schnorr (1971) and others have continued his work. The air is clear for examination of the substantive claim that probabilities can be interpreted in physical terms as limiting relative frequencies of attributes in particular infinite sequences of events.

The simplest examples are provided by binary stochastic processes such as coin-tossing. Here, Mises conceives of an unknown member, h, of the set of all functions from the positive integers to the set $\{0, 1\}$ as representing *the* sequence of outputs that the process would produce if it ran forever. He then identifies the physical probabilities of attributes as the limiting relative frequencies of those attributes in that sequence, e.g. in the case of tosses of a particular coin, h is defined by the condition

(1) $h(i) = 1$ *iff the i'th toss (if there were one) would yield a head,*

and the probability of the attribute *head* is defined,

(2) $p(head) = \lim\limits_{n \to \infty} \dfrac{1}{n} \sum\limits_{i=1}^{n} h(i).$

Both parts of this definition are essential to Mises' attempt to interpret $p(head)$ as a physical magnitude.

In their algorithmic theory of randomness for infinite sequences, Martin-Löf, Schnorr, *et al.* have provided satisfactory abstract models within which part (2) of the definition makes mathematical sense. Thus, Martin-Löf (1968) proposes a model in which Mises' irregular collectives are represented by the set of all functions h that belong to all sets of Lebesgue measure 1 that are definable in the constructive infinitary propositional calculus, e.g. the set of sequences for which $p(\text{head}) = \frac{1}{2}$ in (2). In proving that the intersection of all such sets has measure 1, he shows that his definition escapes the fate of von Mises' (according to which there would be no random sequences) and yields the desired result, that 'almost' all infinite binary sequences are random. The condition $p(head) = \frac{1}{2}$ is inessential: The same approach works for Bernoulli trials with any probability of *head* on each.

But the brilliance of this abstract model of Bernoulli trials is far from showing how probability is connected with physical reality: Rather, it deepens the obscurity of Mises' condition (1), which purports to provide that connection. For most coins are never tossed, and those that are, are never tossed more than finite numbers of times. No infinite sequence of physical events determines the function h of (2): For all but a finite number of values of 'i', the clause following 'iff' in (1) must be taken quite seriously as a counterfactual conditional. But unless the coin has two heads or two tails (or the process is otherwise rigged), there is no telling whether the coin would have landed head up on a toss that never takes place. That's what probability is all about.

A coin is tossed 20 times in its entire career. Would it have landed head up if it had been tossed once more? We tend to feel that there must be a truth of the matter, that could have been ascertained by performing a simple physical experiment, viz., toss the coin once more, and see how it lands. But there is no truth of the matter if there is no 21st toss. The impression that there *is* a truth of the matter arises through the analogy between (a) extending a series of tosses of a coin, and (b) extending a series of measurements of a physical parameter, e.g. mass of a certain planet. If $p(head)$ is a physical parameter on a par with m(Neptune) then–the argument goes–(a) really is just like (b). But the analogy is a false one because while Neptune exists, and has a mass whether or not we measure it to a certain accuracy, the 21st toss of a coin that is only tossed 20 times does not exist, and has no outcome: Neither *head* nor *tail*. A truer analogy would compare p(head) with $m(x)$ where x is a nonexistent planet, e.g. the 10th from the Sun. Mises defines $p(head)$ as the limiting relative frequency of heads in an infinite sequence that has no physical existence. If one could and did toss the coin forever (without changing its physical characteristics) one would have brought such a sequence into physical existence, just as one would have brought an extra planet into existence by suitable godlike feats, if one were capable of them and carried them out. But in the real world, neither the sequence nor the planet exist, and the one is as far from having a limiting relative frequency of heads as the other is from having a mass.

Granted: There is a telling difference between the two cases. In the case of the nonexistent 10th planet we are at a loss to say what its mass would be if it had one, while in the case of the coin that is tossed just 20 times we are ready enough to name a probability for heads. If the coin is a short cylinder with differently marked ends and homogeneous mass distribution, we are confident that heads have probability $\frac{1}{2}$. But this difference tells against Mises: It identifies the probability of heads as a physical parameter of the coin, whether or not it is ever tossed, in terms of which we explain and predict actual finite sequences of events–directly, and not by reference to a nonexistent infinite sequence of tosses. It is because the probability of heads is $\frac{1}{2}$ that we grant: If the coin *were* tossed ad infinitum without changing its physical characteristics, the limiting relative frequency of heads would be $\frac{1}{2}$. But since there is no infinite sequence of tosses, 'its' characteristics cannot explain why heads have probability $\frac{1}{2}$.

2. IRREGULAR FINITE SEQUENCES

In the 1960's, Kolmogorov and others (Chaitin, Solomonoff) founded a theory of algorithmic complexity of finite sequences that sheds fresh light on probability. In showing that the sequences irregular in Kolmogorov's sense are those that pass a certain universal test for randomness, Martin-Löf (1966) provided an alternative definition of irregularity that he was able to extend quite naturally to the case of infinite sequences. In deprecating the foundational importance of the infinite case, I am far from denying the importance and foundational relevance of the finite case as treated by Kolmogorov, Martin-Löf, and the others. What I do wish to deny is that by continuity with the finite case, or by mathematical infection from it, the infinite case gets the importance it would have if ours were a world in which each Bernoulli process went on forever (and in which each Markov process, infinitely replicated, went on forever). To get a sense of the importance and autonomy of the finite case, let us review it briefly.

Tables of 'random numbers' are long, irregular sequences of digits—binary digits, let us suppose. The easiest and surest way to generate such sequences is by Bernoulli processes with equiprobability for the two outcomes on each trial, e.g. by repeated tosses of a coin, with heads recorded as '1's and tails as '0's. In principle, such a process could yield a table of a million '1's, but in practice, no one would buy such a table or give it shelf space.

Why? Well, why spend the money? The table is utterly regular, the relevant rule being, 'Write 1 000 000 ones'. It is not only cheaper but easier to use that rule in your head than to buy and consult the table. *Moral*: We use 'equiBernoulli' processes to generate tables of 'random numbers' not because we have use for the outputs of such processes no matter what they may prove to be, but because we expect such outputs to be irregular, and it is irregularity of the sequence that we seek, irrespective of its provenance.

Kolmogorov (1963) pointed to *incompressibility* as a definitive characteristic of irregularity of finite sequences. Thus, a string of 1 000 000 '1's is compressible, for the rule 'Write 1 000 000 ones' would be only some 100 binary digits long if letters, digits, and spaces were coded in some fairly simple way as blocks of binary digits. In detail, questions of compressibility are relative to (1) choice of one out of the infinity of universal systems of algorithms or programming schemes for generating binary sequences, and to (2) choice of one out of the infinity of measures of complexity of algorithms belonging to the same universal system–let us say, via length of representation in one of the infinity of effective binary coding schemes. Once these choices have been made, we have the means to define the *irregular* ('random') finite sequences as those *about as long as the shortest binary coded algorithms that generate them*. If we define the *algorithmic complexity* of a finite binary sequence as *the length of the shortest binary coded algorithms that generate it*, then the irregular sequences are those whose lengths are approximately equal to their algorithmic complexities.

Locally (i.e. for each particular sequence) the relativity of algorithmic complexity to choices (1) and (2) is problematical (cf. Goodman's (1955) 'grue' paradox), but globally its effect is negligible, for if k_1 and k_2 are two particular measures of algorithmic complexity, there will be a finite bound on the absolute differences between $k_1(s)$ and $k_2(s)$ as 's' ranges over all finite binary sequences. Thus, one proves that the percentage of irregular sequences among all sequences of the same length approaches 100 as the length of the sequences increases without bound: *For large n, practically all sequences of length n are irregular.*

Why do we turn to equiBernoulli processes as sources of irregular finite sequences? Kolmogorov's theory provides a clear answer, as follows. (1) For such processes, all output sequences of length n have probability 2^{-n}. (2) For large n, practically all sequences of length n are irregular. Therefore: (3) For large n, the probability is practically 1 that the output of such a process will be irregular. Then devices lie ready to hand that, with practical certainty, generate long irregular sequences. But mathematical certainty about irregularity is far more difficult to attain: (4) For universal systems of algorithms, the halting problem is unsolvable, and therefore there is no effective test for irregularity of finite sequences. In principle, one might nevertheless be able to prove that particular finite sequences are irregular, but in practice we do well to rest content with high probability.

3. MIXED BAYESIANISM

Suppose that a coin is tossed 40 times, and the process yields nothing but heads. There is a dim argument to the effect that this should not surprise us, for the sequence of 40 heads is no less probable than any other sequence of that length, be it ever so irregular. Of course, this argument must be wrong if we rightly see in such an output compelling evidence that the source was not as we had supposed it to be, e.g. if we see the output as overwhelming evidence that the source, far from being equiBernoullian, is one that yields heads with probability 1 on each toss. But what is the rationale behind this sensible view of the evidence? Here I give a mixed Bayesian answer to this question – 'mixed' in the sense that while statistical hypotheses about the source are treated objectivistically (as hypotheses about physical magnitudes), probabilities of those hypotheses are treated judgementalistically ('subjectivistically ').

(1) Consciously or not, we do or should entertain various hypotheses H_1, H_2, ... about the source, where (in the present example) initially we judge H_1 (the equiBernoullian hypothesis) to be overwhelmingly more probable than H_2 (the hypothesis that heads have probability 1 on each toss), in some such sense as this:

$$\frac{p(H_1)}{p(H_2)} = 2^{20} \approx 1\ 000\ 000.$$

(2) After seeing the output sequence and so verifying the evidence-statement $E = $ 'The output is a string of 40 heads', we revise our judgement via the probability calculus, changing our degree of belief in each hypothesis H from its prior value, $p(H)$, to its posterior value,

$$p(H/E) = p(E/H) \times \frac{p(H)}{p(E)} \qquad \text{(Bayes' Theorem).}$$

so that now H_1 is overwhelmingly less probable than H_2, in the sense that

$$\frac{p(H_1/E)}{p(H_2/E)} = \frac{p(E/H_1)p(H_1)}{p(E/H_2)p(H_2)} = 2^{20}2^{-40} \approx 0.000\ 001.$$

In this Bayesian answer, the probabilities of the hypotheses are 'subjective' in the sense that they are degrees of belief, which need not be 'subjective' in the sense of being ill-founded, arbitrary, or idiosyncratic. But the statistical hypotheses H_1 and H_2 themselves are treated objectivistically. Some Bayesians–notably, DeFinetti–would treat all probabilities as degrees of belief, and others would treat all of them objectivistically. The mixed position represented here is the commonsensical version of Bayesianism that Bayesian extremists must explain away or reproduce within their own terms of reference.

'Bayesians' are so called because of their willingness to use Bayes' theorem in cases where most thoroughgoing objectivists would reject as senseless the prior probabilities $p(H)$ and $p(E)$ of evidence and hypothesis that appear in it. The affinity of Bayesianism with 'subjectivism' (judgementalism) derives from the fact that we may have broadly shared judgements in the form of degrees of belief in H and E even in cases like the present example, where prior inspection of the coin is supposed to have led us to think the equiBernoullian hypothesis overwhelmingly more probable than the other, but where we envisage no definite stochastic process of which the coin is the product–a process of which the ratio of *physical* probabilities would be $p(H_1)/p(H_2) \approx 1\,000\,000$. Pure objectivists who would be Bayesian must envisage some such higher-level process, and treat the prior probability function p as the probability law of that process. Thus, commonsense objectivists sometimes speak (without conviction) of urns containing assortments of coins, some normal, some bent, some two-headed, etc., out of one of which the coin actually used is imagined to have been drawn. In that vision, $p(H_1)$ is the proportion of normal coins in the urn.

Observe that where the 'subject' thinks he knows the objective probability of an event (e.g. the event that all 40 tosses yield heads) and thinks he knows nothing else that bears on the matter, (e.g., perhaps, that the first toss yielded a tail!), he will adopt what he takes to be the objective probability as his degree of belief. Then 'subjective' does not mean *whimsical*. To call a probability 'subjective' is simply to say that it is somebody's degree of belief. One does not thereby deny that the belief has a sound objective basis. Furthermore, the 'events' to which subjective probabilities can be attributed need have no special character (e.g. 'unique', weird, etc.), for they are simply the events concerning which people can have degrees of belief, viz., all events whatever. (These remarks are directed in part to the comments on subjective probability in Schnorr (1971, p. 10).)

4. NONFREQUENTIST OBJECTIVISM

Frequencies are important: The laws of large numbers tell us why, e.g. they tell us that in stationary binary processes, the relative frequency of 'success' will in all probability be very close to the probability of success on the separate trials. Notice that here, the notion of probability appears along with that of relative frequency in the formulation of the law itself. (The notion of probability appears as well in the definition of 'stationary', viz., invariance of probability of specified outcomes on specified trials, under translation of trials.) Frequentism is a doomed attempt to define probability in such a way as to turn the laws of large numbers into tautologies.

The lure of von Mises' program lies in its goal, of providing a uniform, general definition of probability as a physical parameter–a definition that can be applied prior to the scientific discoveries that reveal the detailed physical determinants of stochastic processes, as e.g. the discoveries by Mendel, Crick, and many others revealing the mechanisms underlying the mass phenomena encountered in genetics. Mises sought to found probability as an independent science, on the basis of imaginary infinite sequences of events. Taxed with the unreality of those foundations, he replied that they are as real as the foundations of physics: To measure the physical parameter prob(*head*) to a desired accuracy it suffices to toss the coin often enough, for prob(*head*) is the limit of such a sequence of measurements just as surely as m(*Neptune*) is the limit of another sequence of measurements. Shall we hold the foundations of probability to a higher standard of physical reality than that to which we hold physics itself?

Surely not; but here, Mises holds physics itself to a remarkably low standard of reality, i.e. essentially, the idealist standard to which Bishop Berkeley held it: *Esse percipi est.* The suggestion is that the mass of Neptune exists to the extent to which we measure it, just as the sequence of outcomes exists to the extent to which we toss the coin. As was suggested in Section 1, the limiting relative frequency of heads in the 'ideal' (i.e. nonexistent) infinite sequence of tosses is more properly compared with the mass of some nonexistent planet, e.g. the 10th from the Sun.

But if probabilities are not limiting relative frequencies, what are they? If there is no uniform, general definition of probability that is independent of other scientific inquiries, how shall we define probability as an objective magnitude? I would answer these questions as follows.

The physical determinants of probabilities will vary from class to class of cases; there is no telling *a priori* what they will prove to be. In the case of die-casting, the experience of gamblers and tricksters joins with physical and physiological theory to point to the shape, mass distribution, and (most important) markings of the die itself, as the determinants of the probabilities of the possible outcomes on each toss, and these considerations also join to say that different tosses are probabilistically independent. The case is similar for coin-tossing (where the point about markings is that there *are* two-headed coins about). In lotteries, by design, the determinants are the numbers of tickets of each sort (or the numbers of balls of different colors in the urn), but design is not enough: Empirical and theoretical inquiry may show the design to have been defective, e.g. because the balls of one color share a palpably distinct texture. As with games of chance, so with social, biological and physical probabilities, but even more so: We look to experience, informed with theory, to identify the objective determinants of the probability laws of types of stochastic processes.

The easiest cases are lotteries and urn processes. There, we identify objective statistical hypotheses with the makeup of (say) the urn and, by a happy accident, the probability of drawing a ball of a certain color is numerically equal to the proportion of balls of that color in the urn. In practically all other cases, such a numerical coincidence is lacking. The 'classical' view tried to generalize that coincidence to all stochastic processes. The frequentist view tries to generalize a different coincidence – one that is probable where the law of large numbers holds. On a nonfrequentist objectivistic view, one must face the fact that typically, no such coincidence will be forthcoming – not uniformly in all cases, and not even differentially, on a case-by-case basis. Still, we are often in a position where we can be fairly sure that the relevant determinants, difficult as they may be to describe explicitly and in detail, are the same in two processes, as when we ascertain that two coins were cast in the same mold under similar conditions: Believing that the determinants are shape, mass distribution, and markings, and having good reason to think that these determinants were determined in the same way for the two coins, we have good reason to think that the same probability law will govern the two processes of tossing them – even though we are at a loss to specify the common shape or the common mass distribution except ostensively.

No pure objectivist, I think it important to use judgemental probabilities, e.g. as illustrated in §3 (in an extreme, simplified example).

The present suggestion is that the objective statistical hypotheses to which judgemental probabilities are attributed in such cases will be hypotheses about various kinds of physical magnitudes, which we shall seldom be in a position to specify explicitly and in detail, but which we can often identify ostensively, well enough for our purposes, once we understand what the *kinds* of magnitudes are, that determine the process at hand – kinds like shape, mass distribution, and marking.

This is a far cry from Mises' uniform, general identification of probability with a particular physical magnitude, found in all cases; but as we have noticed, that magnitude does not exist.

Princeton University

BIBLIOGRAPHY

Chaitin, G. J.: 1966, 'On the Length of Programs for Computing Finite Binary Sequences', *J. Assn. Computing Machinery* **13**, 547–569.
Chaitin, G. J.: 1969, 'On the Length of Programs for Computing Finite Binary Sequences: Statistical Considerations', *J. Assn. Computing Machinery* **16**, 145–159.
DeFinetti, B.: 1974, 1975, *Theory of Probability*, Wiley, 2 vols.
Goodman, N.: 1955, *Fact, Fiction, and Forecast*, Harvard.
Kolmogorov, A. N.: 1963, 'On Tables of Random Numbers', *Sankhyā*, Ser. A **25**, 369–376.
Kolmogorov, A. N.: 1965, 'Three Approaches to Definition of the Concept of Information Content' (Russian), *Probl. Peredači Inform.* **1**, 3–11.
Martin-Löf, P.: 1966, 'The Definition of Random Sequences', *Information and Control* **6**, 602–619.
Martin-Löf, P.: 1970, 'On the Notion of Randomness' in A. Kino *et al.* (eds.), *Intuitionism and Proof Theory* (Proc. of Summer Conf., Buffalo, N.Y., 1968), North-Holland.
Mises, R. v.: 1919, 'Grundlagen der Wahrscheinlichkeitstheorie', *Math. Z.* **5**, 52–99.
Mises, R. v.: 1928, 1951, *Probability, Statistics, and Truth* (2nd revised English ed., Macmillan, 1957).
Mises, R. v.: 1964, *Mathematical Theory of Probability and Statistics*, Academic Press.
Schnorr, C. P.: 1971, *Zufälligkeit und Wahrscheinlichkeit*, Springer Lecture Notes in Mathematics **218**.
Solomonoff, R. J.: 1964, 'A Formal Theory of Inductive Inference', *Information and Control* **7**, 1–22.

V

FOUNDATIONAL PROBLEMS
IN LINGUISTICS

ASA KASHER

FOUNDATIONS OF PHILOSOPHICAL PRAGMATICS*

I

Pragmatics has suffered from various debilitating diseases. In this paper I would like to point out one of them and to propose a certain cure; I shall do this by propounding new goals for pragmatical theories and by presenting some major ingredients of any theory which will effect those goals with a view to provide a remedy for a disease I would like to dub 'Pragmaticitis Separatosis'.

To get some idea of the problem, let us take a quick look at Rudolf Carnap's way of disentangling his book *Introduction to Semantics* from any pragmatical complication. He adopts a characterization of pragmatics according to which an investigation is assigned to that field of language study if an explicit reference is made in it 'to the user of a language'.[1] Now, Carnap's trick is to keep the language separate from the user – a language-system is defined by syntactical formation-rules and semantical rules of interpretation, without having any recourse to the language users. Thus, pragmatical rules are either trivial generalizations, describing the prevalent use of a certain language-system in a certain community, or else statements concerning relationships between linguistic behaviour and some non-linguistic behaviour:

'Whenever the people utter a sentence of the form '⋯ ist kalt', where '⋯' is the name of a thing, they intend to assert that the thing in question is cold';
'when using the name 'titisee' the people often think of plenty of fish and good meals.'[2]

Carnap's separatism stems, undoubtedly, from the special attention he paid to constructed languages, where linguistic systems are defined in terms of syntactical and semantical rules. However, when natural languages are under consideration, sequestering the language from its typical user is in want of justification.[3]

The official view of pragmatics among linguists has also been a kind of separatism. J. J. Katz and J. A. Fodor produced in 1963 an invalid argument to the effect that linguistic theory equals grammar plus

Butts and Hintikka (eds.), *Basic Problems in Methodology and Linguistics*, 225–242.
Copyright © 1977 by D. Reidel Publishing Company, Dordrecht-Holland. All Rights Reserved.

semantic theory[4] and recently Katz and T. D. Langendoen have assigned to pragmatics the goal of 'a performance theory at the semantic level' – it is 'a system of rules that specifies how contextual factors interact with grammatical structure to determine an utterance meaning for each token of a sentence type'.[5] Some other linguists, mainly of the Generative Semantics wave, have pointed out interesting phenomena of language-use but have failed to provide an appropriate theory for their presentation and explanation.[6] Finally, Richard Montague assigned some of his formal works to the field of pragmatics, but actually his pragmatical frameworks are of only a slight pragmatical interest, being trivial extensions of his extremely interesting semantical frameworks.[7] Richmond Thomason has aptly labelled this kind of work 'indexical semantics'.[8]

I would like to give countenance to an anti-separatist approach to pragmatics. Whereas separatism takes it for granted that language is completely defined in terms of syntactical structures and semantical relations, the starting-point of the unionist view I endorse is the conviction that no tool is mastered without a thorough grasp of its standard uses.

II

A theory about a tool will be explanatorily inadequate if it fails to specify the *constitutive functions* of the tool. Indeed, abstraction from uses is always possible and is often fruitful, but it involves some partiality which should not lead us astray. Accepting the thesis that no tool can be mastered without an appropriate grasp of its standard uses, and the truism that natural languages are tools, is obviously incompatible with approving most separatist views.[9]

Accordingly, the goal of pragmatics – the ultimate goal of any pragmatical theory – is to specify and explain *the constitutive rules of the human competence to use linguistic means for effecting basic purposes.*

Before elaborating this proposal and applying it, let me mention an analogy. Consider a policeman who is manually operating a system of traffic lights at one intersection. If what he knows about the system is just which button should be pressed in order to turn on any light to any direction, then he has not mastered the system, but only its *syntax*, so to speak. Even if in addition to it he knows what is the '*meaning*' of a red light or a green one, when flashed from his system in any direction, he still has not mastered it. Flashing green lights to all intersecting directions at one and the same time is compatible with whatever he knows about the system, which shows that his understanding of it is partial; syntax and semantics are not enough. As long as he does not know how to facilitate movement of people and vehicles along some street by means of the device of buttons and lamps, he has not mastered the tool of traffic-lights and he does not understand it. Notice that there is still room for a distinction between a policeman that operates his system of traffic-lights well and one who does it badly; one who facilitates traffic poorly by means of a certain device should not be confused with one who does not facilitate traffic at all by means of the same device.

I would like to turn now to a brief elucidation of what I have proposed to be the subject matter of pragmatics. By confining pragmatics to a study of *competence* I mean to exclude from it any systematic consideration of independent psychological factors of human behaviour, like memory limitations, which take part in linguistic as well as in non-linguistic behaviour. Using another analogy, *pragmatics is the arithmetic of language-use rather than the computer engineering of it.*

I take it that a competence is defined by a *system of rules*, both internally – in the human mind – and externally – in the theories of linguists and philosophers. If two systems of such rules are not variants of each other, they might constitute different competences. These rules define, in a sense, the standard uses of linguistic means. Hence the rules under consideration are *constitutive* rules of pragmatics rather than regulative ones.[10]

The competence under consideration is that of using linguistic means for certain purposes. Linguistic means are utterances of sentences. This is, of course, a very simple observation, but it hums with complication. For the present purposes suffice it to say that sentences are not just series of inscriptions or of word-types; they are amenable to unambiguous representations on different levels. On one level, for example, they have intonational contours; on another level they have radicals to which truth-conditions are attributed. Notice that sentences are usable only in suitable contexts of utterance. If an utterance is characterized by means of a sentence and an appropriate context, then sentences, contexts and utterances are linguistic means.

Perhaps it is not needless to remark that by talking about linguistic means, in the present context, we neither commit a vicious circle of describing a language in terms of its own means, nor do we commit ourselves to the assumption that the whole family of linguistic means of a certain language is independently definable. Probably, both uses and means will be characterized by the same multiple recursive definition.

Finally, let me explain shortly what are *basic* purposes of linguistic means. Although poems, for example, comprise words and sentences, the formal patterns of sonnets are not treated of in theories of *basic* uses of language, because using sentences for composing a sonnet presupposes that the same sentences are usable for effecting the standard purposes of, say, asserting and asking. Paraphrasing a remark by Donald Davidson,[11] I would say that there must be the literal use of uttering a sentence in a suitable context, if there are other uses. The theory of use deals with the literal purposes. Basic uses are those that do not presuppose other uses of the same means.

III

After setting up a general, ultimate goal for pragmatical theories, some adequacy conditions should be imposed. Two conditions are of prime importance. The first is fairly obvious: the system of rules should be finitely representable. Some reasons for imposing such a restriction on theories of human competences have been detailed by Noam Chomsky and by Donald Davidson,[12] and will not be dwelt upon here.

The second adequacy condition is as follows:

> For every sentence S of natural language L, every context of utterance C, and every ideal speaker[13] Alpha of the language L, the following biconditional should be a true theorem:
>
> *Context C is linguistically appropriate for speaker Alpha to utter in it sentence S of language L if and only if there is a linguistic institution of language L which grants speaker Alpha an institutional role, enabling him to achieve a basic*

*purpose he entertains in context C, by uttering in it sentence
S.*

I shall call this condition 'Convention A' about appropriateness.

There is some correspondence between this convention A and the role it plays in pragmatical theory and convention T, introduced by Tarski and defended by Davidson, and its role in semantical theory.[14] An adequate theory of the linguistic appropriateness relationship between sentences, contexts and speakers, in terms of linguistic institutions, roles and purposes, will constitute an adequate theory of language-use competence. An important distinction between convention T and convention A, is that the former moderates the use-mention distinction, at least for contexts of truth predication, while convention A does not have similar moderating assignments.

The key-concepts of convention A are: linguistic appropriateness, linguistic institution, institutional role, and basic purpose. Formally speaking, lingustic appropriateness is a relation between sentences, speakers and contexts of utterance, but on a deeper level it relates *purposes* people have and the linguistic *means* they employ for effecting them.

Now, the linguistic potential of an act of uttering a sentence in a context arises from a related linguistic *institution*, that is, a system of non-natural rules that govern that kind of activity. An institution is not just a system of constitutive rules; institutions have three additional properties that should be mentioned. First, institutions are means of *coordination* between diverse goals of different persons. This is a common observation and I shall employ it for some purpose later.

Secondly, institutions assign *roles* to diverse people under diverse circumstances. Usually, rules of roles determine requirements a person has to fulfill under circumstances of a certain type, in order to be able to play the role under consideration. Some of the rules which John Searle has put forward for the institution of promising are clearly role-oriented.[15] The contribution of the context of utterance to the working of linguistic institutions is, thus, not confined to providing values for some obvious indices; it also determines who is in a position to play some role in some linguistic institution.

Thirdly, by playing their institutional roles, people *institute* new facts, rendering some acts, like inscribing or uttering sounds, meaningful and useful beyond their natural properties and potential. The fact that proposition P was expressed by speaker Alpha who uttered sentence S in context C, is a typical institutional fact.

Convention A points out some required conceptual resources of adequate pragmatical theories. Convention A and the other adequacy condition (finite representability) seem to justify labelling any pragmatical theory that satisfies both of them '*generative*'. We do not intend to blur a distinction between philosophy and linguistics, *viz.* that between foundational proposals (e.g. convention A or any conceptual analysis of the concepts of linguistic institution, role and purpose, which are philosophical in nature) and empirical theories (e.g. a grammar or a recursive specification of properties of sentences and contexts, which are of a linguistic nature). However, bridging some gaps between philosophy of language and linguistics does not seem to me to be pointless.[16]

IV

Having outlined general goals and adequacy conditions for pragmatical theories, I will turn now to a discussion of two kinds of constitutive rules of language use.

Since the subject matter of pragmatics is to my mind the system or systems of constitutive rules of the human competence to use certain *means* for achieving certain *purposes*, it seems only too natural to look for constitutive rules that will reflect the share taken in pragmatical theory by the most general conceptual framework of *uses*. I would like to suggest that such a constitutive role is played by some principles of *rationality*. Of course, nobody is a perfectly rational agent, but when our competence to use linguistic means to attain basic goals is under consideration, rather than our actual performance, we assume that a weak version of the principle of effective means is followed. Trying to apply the usual, strong version of that rationality principle to the linguistic case, we face the following requirement:

> Given a desired basic purpose, the ideal speaker chooses that linguistic action which most effectively and at least cost attains that purpose, *ceteris paribus*.[17]

To see that this is *not* a constitutive rule of the linguistic use competence, consider the case of a speaker whose speech is effective but not terse; he attains his basic goals effectively but not always at least cost. Should he be excluded from membership in the community of language users? Of course not.

However, we do require that given a desired basic purpose, the ideal speaker chooses that linguistic action which – *to the best of his belief* – most effectively and at least cost attains that basic purpose, *ceteris paribus*. If someone believes that he cannot attain the basic purpose he entertains at a certain context *at least cost* without uttering a certain sentence, then something is wrong with his linguistic competence if he utters a completely different sentence, deliberately and for no overriding reasons. It is important to notice that the cost of deliberation is always of significance. Given the amounts of, say, time and energy that the speaker is willing, under certain circumstances, to invest in deliberation, he is supposed to choose what he believes to be the best sentence that he can formulate for attaining the present basic purpose he entertains without indulging in a process of deliberation which requires more than the given amounts of time and energy. This is why most of the time most of us do not sound terse. In daily speech, we are very reluctant to spend more than a second or so in trying to find a better way of expressing what we have in mind; this is not so, indeed, under diplomatic and other formal circumstances.[18]

The strong version of the applied principle of effective means and the weak, doxastic version of it, introduce two different standards for linguistic action. The weaker version sorts out one competence from other competences and capacities, while the stronger, objective version grades language users by excellence. Thus, the distinction between constitutive rules and strategies is well marked here.

A discussion of rationality principles might have sounded too abstract to be of any value to pragmatical theory. It is, however, possible to dismiss that impression by recalling H. P. Grice's theory of implicatures, as presented in the various parts of his William James lectures.[19] Grice's theory proposes to explain a host of linguistic intuitions we have concerning what is conveyed implicitly by some utterances, and tries to do it by applying a principle of cooperation and several conversational maxims. The former principle is that of making conversational contributions such as required, at the stages at which they occur, by the accepted purposes or directions of the talk exchanges in which the conversational contributors are engaged. This principle allegedly grounds some less general 'super-maxims'. The super-maxim of *quantity* commends not making a conversational contribution less or more informative than is required; the super-maxim of *quality* is that of trying to make conversational contributions ones that are true; the super-maxim of *relation* commends relevancy, and that of *manner* commends perspicuity.

It seems to me that the principle of cooperation is wrong and that the super-maxims are not derivable from it anyway. But rather than indulging in a detailed review of Grice's main points, which I have done elsewhere,[20] let me just hint at some difficulties and point out better foundations on which Grice's extremely powerful explanatory framework of implicature-production should rest.

The principle of cooperation is at once too strong and too weak. On the one hand, it relies on the assumption that in all stages of any conversation, it is always possible to identify the joint purpose or direction of it. Now, do Socrates and the slave-boy from among Meno's companions, who is learning some geometry from him, share any purpose? Or do Socrates and Crito, who are engaged in a conversation in which each of them tries to convince the other that he is in error, share any global or local conversational purpose? The principle is too strong since it rests on an assumption of cooperation and shared purposes rather than on the weaker assumption of coordination and independent purposes. On the other hand, the principle of cooperation commends striving for some desired ends but it does not impose restrictions on the *ways* to be taken for effecting these ends. Since all the super-maxims are tantamount to such restrictions, the principle does not ground them.

The principle of cooperation should be replaced by the following *principle of rational coordination:*

> Given a desired basic purpose, the ideal speaker chooses that linguistic action which, he believes, most effectively and at least cost attains that purpose, *ceteris paribus.*

I shall now elaborate a little on two aspects of this principle, namely – rationality and coordination.

The principle of rational coordination has four general consequences, concerning ways of attaining purposes. Each of Grice's four super-maxims is just a particular case of a corresponding consequence. An obvious consequence of the principle of rational coordination is that means should not be used for achieving given desired ends, more or less than is required for the achievement of these ends, *ceteris paribus*. Grice's super-maxim of quantity is evidently a consequence of that ultra-maxim of rationality, where information is taken to be a measure of linguistic means. A second ultra-maxim, drawn from the principle of rational coordination, commends trying to achieve given desired ends by *standard* use of the available means, *ceteris paribus*. This is not an obvious conclusion of that principle, but here I would like just to note down that Grice's super-maxim of quality follows from this ultra-maxim, using David Lewis's way of defending the convention of truthfulness.[21]

A third ultra-maxim is as follows: At every stage on your way to the achievement of your desired ends, consider the means being used by other agents to achieve their goals, when you determine the manner of your using your means; moreover, prefer using your means in a manner which is likely to help the progress of other ones in their ways to the achievement of their desired ends over any other way of using your means, *ceteris paribus*. Without attempting here a derivation of this ultra-maxim, I would offer it as an explication of Grice's super-maxim of relation, which is both suggestive and vague in its original form of commending relevancy.[22]

Finally, the following ultra-maxim, derivable from the same principle, entails Grice's super-maxim of manner which commends perspicuity: Give preference to use of available means which leads you directly to your desired ends, over use of those means which leads you to situations wherein achievement of your ends is just a possible, non-standard result, *ceteris paribus*.

The replacement of the principle of cooperation by the principle of rational coordination, and the derivation of the super-maxims of Grice from our ultra-maxims, require one amendment in Grice's ingenious method of deriving implicatures. Instead of assuming that the speaker is observing the principle of cooperation or the super-maxims, we assume that the speaker's basic ends and beliefs justify his linguistic behaviour. In other words, we use the principle of rational coordination as a premiss in every derivation of implicatures, unless there is reason to reject this assumption. For practical reasons it is, indeed, sufficient sometimes to assume that certain ultra-maxims, super-maxims or even particular maxims are observed by the speaker. The generative power of Grice's theory of implicatures is thus left intact. Let me, however, mention very briefly two advantages of the presently proposed theory over Grice's one.

First, consider any case of what Grice calls 'flouting a maxim' by uttering, for example, a tautology. Grice's theory explains why the speaker uttered one tautology rather than any other one, but it does not explain why the implicit way of conveying opinions, by means of implicatures, has been preferred by the speaker to an explicit way of conveying his point by means of frank assertions. A rational reconstruction of the speaker's linguistic behaviour has to explain this preference, but Grice's theory leaves it out of view.

Secondly, let me mention a peculiar species of speech acts, namely silence acts. Consider, for instance, my deliberate refraining from answering a question directed to me. My silence implicates some information, as shown by using the principle of rational coordination. Similar implicatures do not seem to be derivable from the principle of cooperation and the super-maxims.

Having outlined a few ideas concerning one aspect of the principle of rational coordination – that of rationality – I would like to say now a few words about another aspect of it, *viz.* coordination.

Coordination, in a broad sense of the word, is the general purpose served by institutions, but many acts performed under diverse institutional circumstances do not serve that purpose directly. Non-verbal ceremonies, for example, are not satisfactory for any coordination purpose, but they establish institutional facts and preserve frozen traditions which in turn do enhance coordination. However, linguistic institutions bear the outstanding characteristic of comprising only actions which serve directly basic coordination purposes. Every speech act conveys a preference on the part of the speaker of some possible worlds over some other ones. The preference-expressions involved take standard forms, depending on the so-called 'mood' or 'force' of the uttered sentence; which possible worlds are preferred to which ones are determined in part by the so-called 'radical' or 'propositional part' of the sentence, and by all the presuppositions involved. I developed these ideas in much detail elsewhere, so I shall leave it at that.[23]

V

Thus far I have discussed only one, special kind of constitutive rule of language-use, *viz.* rules that reflect the fact that linguistic behaviour is purposeful and, in a sense, rational. I have also mentioned a family of constitutive rules pertaining to a particular feature of the linguistic tool, namely that all speech acts are meant to lead directly toward coordination. The next kind of constitutive rule of language-use is of much lesser generality, appertaining to the uses of certain expressions in some natural languages.

Consider the ambiguity of sentences like 'Mary wants to marry a mayor' or 'A picture is missing from the gallery'. Philosophers have discussed it under the label of 'the referential/attributive dichotomy', which is related to the specific/non-specific dichotomy of the linguists, but since attention has been paid mainly to ambiguities related to definite rather than indefinite expressions, the theories provided by philosophers or linguists do not seem to be adequate.

A promising way to obtain an explanation of that ambiguity is to try to accomodate the formal framework of Jaakko Hintikka's language-games with rules that would apply to indefinite expressions, under all their readings.[24]

Those language-games may be regarded as being played by the speaker, who utters the sentence under consideration in an appropriate context of utterance, and a listener. The former tries to provide support for what he said, while the latter tries to subject it to extreme criticism, so as to see whether there is some flaw in the speaker's words that should prevent the listener from accepting them. Under this interpretation, which should not be confused with the dialogical games of Lorenzen and others,[25] the rules of Hintikka's language-games are constitutive rules of the linguistic institution of backing up and criticizing, obviously related to the rationality ultra-maxim of quality.

Recall that the games proceed in line with the structures of the sentences under consideration; with every move the players go on from a given formula and a given possible world in which it is claimed to hold, to a subformula and a related possible world, possibly identical to the given one. The moves of the players are choices of subformulae, individuals or possible worlds, according to the local interests of the players – supporting or refuting a given formula under given conditions.

To mention just one rule: when a universal formula is given, the move belongs to the player whose role is to refute it. He chooses an individual, from an appropriate domain, and ascribes to it a suitable proper name, whether established or new. Now, the quantifier and the variable attached to it in the given formula are dropped from it and every occurrence of the variable within the matrix of the formula is replaced by the chosen proper name. The resulting formula is that given in the next move; the given possible world will serve for the next step of the game as well. All the choices and replacements are done openly: both players know exactly what is given and the player who makes the choice at a move informs his opponent of the identity of the chosen individual, possible world or subformula, without any sort of disguise. Put in game-theoretical terms, Hintikka's language-games for classical connectives, quantifiers and operators are *games of perfect information*.

In order to equip that language-game framework with rules for all readings of indefinite expressions, it seems necessary to waive that condition of perfect information. When indefinite expressions are major constituents of the given formula individuals are selected and given proper names, but these operations are not performed openly, and only partial information is shared by all participants.

For one reading of the sentence 'Mary wants to marry a mayor', according to which it is true of a certain mayor whom the speaker has in mind that Mary wants to marry him, the speaker introduces a new name, say 'Beta', for that certain mayor, and goes on playing with the formula roughly corresponding to the sentence 'Of Beta it is true that Mary wants to marry him'. The listener goes on playing with the same formula, but he has not been informed about the identity of Beta. What he does know is that Beta is a mayor familiar in a way to the speaker.[26]

According to another reading of the sentence, it is not the speaker but Mary who has a certain mayor in mind. Thus the game continues with a formula roughly corresponding to the expression 'Mary wants of Gamma that she is married to him', where 'Gamma' is a new proper name for that mayor. The selection is done this time by Mary, so to speak, and both the speaker and the listener are only partly informed about his identity; they know he is a mayor familiar in a way to Mary.[27]

The gist of the example of specific indefinite expressions is that the use of these expressions is governed by special constitutive rules that involve partial descriptions of individuals which are identifiable through an established proper name or a uniquely identifying description, appropriately available. These are non-trivial rules that belong to the system of constitutive rules of a human competence, related to the use of *certain* linguistic means. Similar cases govern the uses of some tenses, nominalizations, comparatives and terms to which Hilary Putnam's principle of the division of linguistic labour applies.[28] All these cases involve in one way or another use of partial information.

There are some important theoretical lessons to be learned from a scrutiny of these examples. Let me mention most briefly three theoretical problems on which some light is thus shed. First, *the speaker/listener symmetry*. Prevalent among theoreticians is the tendency to relegate any non-trivial distinction between speaker and listener to the psycholinguistic domain of performance. According to this view the same competence is used both in the process of language production and in the process of language understanding. However, there is a certain speaker/listener asymmetry within pragmatics, that is to say within one component of the human linguistic competence.

Secondly, *truth-conditions*. In order to capture the concept of partial description within a formal framework, a generalization of the concept of interpretation is required. Partial interpretations are easily defined and interestingly restricted.

Thirdly, *language learning*. It is obvious that a child acquires syntactical and semantical constitutive rules that are specific for the natural language he learns. It is a fact of some interest that there are also pragmatical rules which are so specific.[29]

VI

I shall conclude this paper by trying to answer an operative question, namely – 'what are pragmatical representations?' The formal branch of our philosophical heritage taught us to put the burden of representation upon formulae and rules which apply to them in one way or another. Blending these two representational elements to maintain some explanatory equilibrium is always a tremendous task; no attempt to defend the proposed blend will be made here.

A theory about a human competence to use linguistic means for attaining basic purposes, should enable us to represent institutional facts concerning sentences, i.e. it should enable us to represent basic purposes, roles, means and uses, and their properties.

Each utterance of a sentence, in an appropriate context of utterance, corresponds to a basic pragmatical formula, composed of a *radical* formula and a *mood* indicator. The radical formula represents the propositional part of the utterance to which truth-conditions apply. It is an interpreted formula in an extended logical language. (An extension is required since radical formulae are not just truth-bearers; they have all kinds of pragmatical duties to fulfil. To mention just one example: the radical formula should represent indexical contributions of the context of utterance conspicuously.)

The mood indicators designate *basic forces* of sentences, e.g. assertoric, imperative, performative. Formally speaking, these indicators are functions that ascribe to radicals characteristic preference relations of some possible worlds over others. These preference relations represent the *basic purposes* of the corresponding speech acts.

Different *roles* played by persons engaged in certain linguistic actions, are duly characterized by systems of *felicity preconditions*. A basic pragmatical formula, i.e. a radical formula and an attached mood indicator, determines both the form and the content of the related system of felicity conditions. (Notice that here is another duty the basic formulae fulfil: simple rules read between their parentheses, so to speak, all the presuppositions involved.[30])

Once roles and basic purposes have been determined, institutional affiliations are established. The mood indicators mark the most basic among these institutions, e.g. imperative and performative. When additional preconditions are admitted, the way to special institutions, which are marked by illocutionary force indicators is clear. Each of those institutions, whether basic or not, might comprise all kinds of rules, like those of Hintikka's language-game of backing and criticizing which is included in the institution of assertion. Some of the non-basic institutions have essential non-linguistic ingredients.[31]

Finally, *implicatures* should be treated within the outlined framework, thus capturing the concepts of linguistic *means* and *uses*, at least in part. Given a radical formula and a mood indicator, standard linguistic rules like transformations and lexical insertions are used to produce any additional linguistic representation of the utterance which is relevant to the derivation of implicatures. For example, intonation contours should be represented because their employment conveys some information systematically. Properties of these representations are extracted to provide premisses for proving rationality-theorems. The conversational ultra-maxims serve as derivation rules for this purpose of deriving implicatures as theorems. The case for deriving *all* implicatures of a certain utterance seems quite dubious.[32]

It was the poet Alexander Pope who unthinkingly expressed the classical goal of linguistic theories, saying: 'Tis not enough no harshness gives offence; The sound must seem an echo to the sense'.[32] I plead for an appropriate formal, generative and institutional amendment: 'Tis not enough no harshness gives abuse; The sound must seem an echo to the use'."

Tel-Aviv University

NOTES

* This paper has benefited from comments by L. Jonathan Cohen, Joel Friedman, Jaakko Hintikka, David Kaplan, Hans Lieb and Stanley Peters. I owe special gratitude to my wife, Naomi, whose conceptions of institutions and rules were very helpful for clarifying some of the basic issues of this paper.

Some of the work related to this paper was supported by the Israel National Academy of Sciences and Humanities, through a grant to Tel-Aviv University. Another part of the work was supported by the Deutsche Forschungsgemeinschaft through a grant to Helmut Schnelle, then at the Technische Universität Berlin.

[1] Carnap (1959, p. 9).

[2] The first example is *Pragm. 1* of Carnap (1939, p. 5) and the second one is part of *Pragm. 2a, ibid.*

[3] Important differences between construction and analysis have often been at least underestimated.

[4] Katz and Fodor (1963, pp. 170–210). See also Bar-Hillel (1967) and Kasher (1970) for some criticisms.

[5] Katz and Langendoen (1975).

[6] See, for example, R. Lakoff (1973).

[7] See Montague (1974) and a methodological discussion of it in Kasher (1975a).

[8] Thomason (1973). The following shares a lot with Stalnaker's approach in his (1974).

[9] The only separatist view which might be compatible with everything that follows from the above mentioned thesis and truism is the one that endorses partial abstractions for special purposes.

[10] For some preliminary clarifications, see Searle (1969, pp. 33ff).

[11] Davidson (1969).

[12] Davidson (1965) and Chomsky (1965, p. 222, n. 2 and the references thereof).

[13] We use the term 'ideal speaker' in the sense given to it in Chomsky's works.

[14] See Davidson (1973).

[15] Searle (1969, pp. 57–61).

[16] Davidson (1974) is interestingly related to this point.

[17] For the role of the *ceteris paribus* clause, here and in sequel, and for a detailed discussion of the related rationality principles, see Richards (1971).

[18] Notice that I am not putting forward here a psychological theory but rather a theory about a competence.

[19] For the published parts, see Grice (1975). For some formalization of the concept of implicature and pragmatical applications, see Kasher (1974). See also Searle (1975).

[20] Kasher (1975c).

[21] Lewis (1969, pp. 148–152, 177–195).

[22] See my (1975c).

[23] Kasher (1974).

[24] For the general framework, see Hintikka (1973) and for a highly interesting extension of it, Hintikka (1974). For the application and extension suggested in the sequel see also my (1975b).

[25] The differences and similarities are discussed by Hintikka in his (1973, pp. 80ff).

[26] Some discussion of the meaning of 'familiar in a way' is included in Kasher and Gabbay (1975).

[27] A third reading of the sentence corresponds to an existential formula related to 'Mary wants there to be a mayor to whom she is married', and is of no particular interest here. I leave out some important details concerning the first two readings as well. See Kasher and Gabbay (1975) for many details.

[28] Putnam (1974).

[29] Although this observation sounds philosophically trivial, it has some significance for the construction of linguistic theories. I owe the latter observation to H.-H. Lieb.

[30] See, for example, my (1973, pp. 390f.). In some points, not all of which are just terminological, that paper is not satisfactory to my mind any more.

[31] At this point I owe much to Cohen (1975).

[32] Here I am indebted to Rich Thomason's paper (1973), though I do not share his pessimism.
[33] I owe this reference to Y. Bronowsky's Hebrew *Essay on Language*.

BIBLIOGRAPHY

Bar-Hillel, Y.: 1967, Review of Katz and Fodor (1964), *Language* **43**, 526–550.
Carnap, R.: 1939, *Foundations of Logic and Mathematics*, The University of Chicago Press, Chicago, Ill.
Carnap, R.: 1959, *Introduction to Semantics*, Harvard University Press, Cambridge, Mass.
Chomsky, N.: 1965, *Aspects of the Theory of Syntax*, MIT Press, Cambridge, Mass.
Cohen, L. J.: 1975, 'Speech Acts', *Current Trends in Linguistics*, Vol. XII (ed. by T. A. Sebeok), Mouton, The Hague.
Davidson, D.: 1965, 'Theories of Meaning and Learnable Languages', *Proceedings of the 2nd International Congress for Logic, Methodology and Philosophy of Science* (ed. by Y. Bar-Hillel), North-Holland, Amsterdam.
Davidson, D.: 1969, 'True to the Facts', *Journal of Philosophy* **LXVI**, 748–764.
Davidson, D.: 1973, 'In Defence of Convention T', *Truth, Syntax and Modality* (ed. by H. Leblanc), North-Holland, Amsterdam and London.
Davidson, D.: 1974, 'On the very Idea of a Conceptual Scheme', *Proceedings and Addresses of the American Philosophical Association*, Vol. XLVII.
Grice, H. P.: 1975, 'Logic and Conversation', *The Logic of Grammar* (ed. by D. Davidson and G. Harman), Dickenson, Encino and Belmont, California.
Hintikka, K. J. J.: 1973, *Logic, Language-games and Information*, Oxford University Press.
Hintikka, K. J. J.: 1974, 'Quantifiers vs Quantification Theory', *Linguistic Inquiry* **5**, 153–177.
Kasher, A.: 1973, 'Logical Forms in Context: Presuppositions and other Preconditions', *The Monist* **57**, 371–395.
Kasher, A.: 1970, 'Semantic Description does not Equal Linguistic Description Minus Grammar', unpublished MS, The Hebrew University of Jerusalem, Applied Logic Branch.
Kasher, A.: 1974, 'Mood Implicatures: A Logical Way of Doing Generative Pragmatics', *Theoretical Linguistics* **1**, 6–38.
Kasher, A.: 1975a, 'The Proper Treatment of Montague Grammars in Natural Logic and Linguistics', *Theoretical Linguistics* **2**, 133–145.
Kasher, A.: 1975b, 'Pragmatical Representations and Language Games: Beyond Extensions and Intensions', *Rudolf Carnap: Logical Empiricist* (ed. by J. Hintikka), Reidel, Dordrecht and Boston.
Kasher, A.: 1975c, 'Conversational Maxims and Rationality', *Language in Focus: Foundations, Methods and Systems*, Essays dedicated to Yehoshua Bar-Hillel (ed. by A. Kasher), D. Reidel, Dordrecht and Boston.
Kasher, A. and Gabbay, D.: 'On the Semantics and the Pragmatics of Specific and Non-specific Indefinite Expressions, I', *Theoretical Linguistics* **3**.

Katz, J. J. and Fodor, J. A.: 1963, 'The Structure of a Semantic Theory', *Language* **39**, 170–210.

Katz, J. J. and Fodor, J. A. (eds.): *The Structure of Language.*

Katz, J. J. and Langendoen, D. T.: 1975, 'Pragmatics and Presuppositions', *Language*, forthcoming.

Lakoff, R.: 1973, 'The Logic of Politeness', *Papers from the 9th Regional Meeting Chicago Linguistic Society.*

Lewis, D.: 1969, *Convention*, Harvard University Press, Cambridge, Mass.

Montague, R.: 1974, *Formal Philosophy*, edited and with an introduction by R. H. Thomason, Yale University Press, New Haven.

Putnam, H.: 1974, "The meaning of 'Meaning' ", MS, Harvard University, Cambridge, Mass.

Richards, D. A. J.: 1971, *A Theory of Reasons for Actions*, Oxford University Press.

Searle, J. R.: 1969, *Speech Acts*, Cambridge University Press.

Searle, J. R.: 1975, 'Indirect Speech Acts', *Speech Acts* (ed. by P. Cole and J. L. Morgan), Academic Press, New York.

Stalnaker, R. C.: 1974, 'Pragmatic Presuppositions', *Semantics and Philosophy* (ed. by M. K. Munitz and P. K. Unger), New York University Press, New York.

Thomason, R. H.: 1973, 'Semantics, Pragmatics, Conversation and Presupposition', MS, Pittsburgh University.

DIETER WUNDERLICH

ON PROBLEMS OF SPEECH ACT THEORY

1. INTRODUCTORY REMARKS

I consider speech act theory to be an extension of the theory of meaning in natural language. Whereas up to now the theory of meaning has almost exclusively been concerned with the meaning of strictly declarative sentences, the aim of speech act theory is to characterize the meaning of non-declarative sentences as well as declarative ones in terms of possible speech acts. As the theory of meaning includes a semantical and a pragmatical part, speech act theory deals with both aspects as well. The difference of the two aspects I consider to be as follows: semantics is concerned with the meaning of sentences in abstraction from context, pragmatics is concerned with the meaning of sentences in contexts. There is no reason to identify speech act theory in general with pragmatics of natural language, since there are kinds of speech acts that can be treated within a proper extended semantics. On the other hand, however, there are also other kinds of speech acts that must be treated by reference to special features of context, i.e. in a pragmatical way.

Let me regard a discourse as consisting of a sequence of utterance acts. In executing an utterance act normally a speech act of a certain kind is performed. We may say that a certain speech act concept of language L is realized, by means of uttering certain forms of L, which, in most cases at least, can be related to certain sentences of L. Each speech act concept of L, i.e. a generic speech act of L, will be identified with the meaning of certain sentences of L. The aim is, then, to construe the speech act concepts of L, each being the meaning of certain sentences of L, as the outcome function for utterances of those sentences within certain contexts. Each speech act concept belongs to a certain illocutionary type. Related with those types are certain purposes of language within interaction processes.

It has often been claimed that languages like our own contain about a thousand or even more different speech acts, including such complex

Butts and Hintikka (eds.), Basic Problems in Methodology and Linguistics, 243–258.
Copyright © 1977 by D. Reidel Publishing Company, Dordrecht-Holland. All Rights Reserved.

ones as e.g. reports, narrations, argumentations and proofs. Prima facie, to be confronted with such a multitude of speech acts is a rather disastrous situation for building up a theory. But the situation is not that hopeless. We can easily see that the claim is mainly based on the existence of a great number of verbs which in fact do not denote single speech acts but merely denote certain aspects, or functions, or modifications of speech acts (verbs like 'answer', 'deny', 'criticize' or 'summon'), or even denote certain discourse types (verbs like 'narrate', and 'argue'). Therefore we will have to reduce the number of fundamental speech acts, e.g. to those that are formally marked by grammatical mood (and not only denoted by verbs), and further to explicate which functions they may have within a discourse of a certain kind, which modifications they may undergo within certain institutions of our society, and which discourse complexes may be built up. One of the main problems will be how the relevant context can be represented within linguistic theory.

I shall proceed as follows. After a short summary of what the theory of meaning does with respect to declarative sentences, I shall argue why most of the hitherto known approaches to non-declarative sentences are not satisfactory. Either they regard non-declarative sentences falsely as a subspecies of the declarative ones, or they treat them in a pragmatical way only, yielding theoretical statements that are weaker than we would like them to be. In the next part of my paper I shall outline a theoretical framework of investigation, in which there will be distinguished semantical and pragmatical aspects of speech acts. To specify the general ideas I shall give a partial semantic definition of two of the interesting illocutionary types, namely of those expressed by the imperative and by the interrogative mood, leaving aside questions of the logical properties of these types. Finally I shall characterize a group of speech acts, which I will call conditional speech acts, that can be handled in a pragmatical way only, on the assumptions of a theory of preferences, together with a notion of practical inference, and a semantical theory of conditionals.

Some other still open problems I cannot discuss here. One problem concerns the so-called Hedged Performatives, i.e. sentences with the performative verb in an embedded clause (like 'we regret to *inform* you that your policy is canceled'). The envisaged solution to this problem will treat the syntactically governing parts of those sentences as expressing a specification of or a commentary on the speech act concept expressed in the dependent clause. Another problem concerns the proper sequencing of speech acts in a discourse. In the light of speech act theory two different models or strategies of explication can be conceived of. The first one uses only the semantic properties of speech act types in constructing kinds of consistent sequences. It may be used to classify discourse types according to their semantical sequence structure. The other one aims at the reconstruction of certain transitions within dialogues. It will crucially use a rather wide notion of practical inference including those kinds of inferences Grice has called Conversational Implicatures. All relevant contextual information will be represented as premises of practical inferences. This strategy may lead to a classification of discourse types according to the types of needed contextual premises. (Cf. Wunderlich, 1975c.)

2. The theory of meaning applied to non-declarative sentences

Under the traditional point of view only one purpose of language has been observed within the theory of meaning, that is the representation of matters of reality, taken in a broad sense. Accordingly the speech act under consideration has been the assertion, expressed by a declarative sentence. Yet if there are no different purposes that have to be distinguished, one can neglect the category of purpose altogether and abstract from the fact that an assertion on the whole is a speech act. Hence, the theory of meaning, restricted to declarative sentences, has dealt with the notion of proposition only.

Formally, a proposition can be represented as a function from possible worlds into truth values. Each proposition divides the total set of possible world alternatives into two subsets; hence, a proposition can also be represented as the set of those possible worlds in which it is true. Semantics studies propositions, and also, which proposition p is expressed simpliciter by a certain declarative sentence s in each of its readings. Indexical semantics studies which proposition p is expressed by a certain s relative to a sequence of fixed features of the context of use c^0 (like speaker, addressee, time of utterance, etc.), given that s contains so-called indexical expressions and expresses simpliciter an open proposition p^0 which is like p except that it has at least one free place to be filled out by a feature of the context of use c^0. Pragmatics studies which proposition q is expressed (or, eventually, which propositions q_1, q_2, etc. are expressed) by a certain s relative to a context of use c, given that s expresses simpliciter p, and that the context of use c can be characterized by certain principles of conversation (which may belong to institutional procedures governing the conversation as a whole) and certain common suppositions of the participants (which may belong to their knowledge, or may be brought about by the foregoing discourse, or by their perception of the surroundings).

As to non-declarative sentences, the various proposals that have been made can be grouped somewhat into two classes. Advocates of the first kind of proposals take the position that non-declaratives can also be analyzed with reference to truth values, whereas advocates of the second kind of proposals take the position that normal utterance situations of non-declaratives can be characterized in pragmatical terms of attitudes of speaker and addressee. I will clarify the respective positions by an example.

According to the first position, the truth conditions of the imperative sentence

(1)   Shut the door.

are identified with, or in a narrow way related to, the truth conditions of one of the following expressions:

(2)   You will shut the door.

(3)   I order you to shut the door.

(4)   You are obliged to shut the door.

(5)   You are requested that, you will shut the door.

as it has similarly been proposed by Cresswell (1973), Lewis (1971), Chellas (1971), and Davidson (1974) respectively. It is true that the cited expressions are partly equivalent with (1) in the sense that they can convey the same order as (1). However, (2) through (5) all are ambiguous, and the proposed treatments just rely on that descriptive reading of (2) through (5) respectively which cannot be given to (1).

(2), in this reading, expresses the statement that it will be the case that the addressee shuts the door (instead of expressing that this ought to be the case). (3), in this reading, expresses the statement that the speaker does make the order to the addressee to shut the door (instead of expressing the order itself). (4), in this reading, expresses the statement that the obligation to shut the door already exists (instead of introducing this obligation). Finally, (5) consists of two sentences. If the first one is true, a certain request already exists; if the second one is true, the request will be complied with. Hence, the arguments against each sentence of (5) taken in isolation correspond to those already given against (3) and (2). These arguments hold even if we take into account a proper reading of the demonstrative 'that' in (5).

It appears that those who advocate this position would say that (2) through (5) all are non-ambiguous but each has different uses, and that we need only different acceptability criteria to distinguish the different uses of the cited sentences. But this amounts to the claim that the illocutionary force of an utterance of anyone of these sentences wholly depends on the context of use. How this claim can be accounted for, and of what kind the acceptability criteria are to be, remains a mystery.

According to the second position, the following statements with respect to a normal utterance situation c of (1) have been made:

(6)   The speaker α intends to get the addressee β to shut the door, by means of getting β to recognize this intention, in virtue of β's knowledge of the rules of language L. (Searle 1969.)

(7)   The speaker α expresses that he wants the addressee β to shut the door. (Searle 1969.)

(8)   The speaker α prefers in c a state of affairs in which the addressee β has shut the door a short while after c over a state of affairs in which β has not shut the door a short while after c, other things being equal. (Kasher, 1974.)

(9)   The addressee β tries to act in such a way that (1) was true in L on the occasion c of its utterance to β. (Lewis, 1969.)

The used notions like 'intend', 'want', 'prefer', and 'try' are pragmatic ones. Since the theoretical statements are concerned with the attitudes of speaker and addressee, they are related to normal contexts of use where speaker and addressee do in fact have these attitudes. However, there seems to be no clear way to characterize the normal context except by just those statements. Hence, the approach bears some circularity in it. (A partial exception is the treatment of Lewis, who calls (9) a truthfulness convention, which we may understand not in a descriptive but normative sense, and who refers to contextual features like the common interest in making it possible for the speaker to control the addressee's actions within a certain range.)

Moreover, since the statements relate to normal contexts of use, they have to be viewed as contextual implicatures of (1), and hence they are weaker than the description of the meaning of (1) itself. We can see this also by the fact that the same implicatures might be drawn from other utterances and even from a non-verbal behavior, provided we enrich the context in an appropriate way. (A partial exception may be the treatment of Searle, if we understand his 'rules of language' in a strong way.)

3. THE EXTENDED THEORY OF MEANING

In order to overcome the mentioned deficiencies I will propose the following treatment.

At first, let me introduce the notion of a neutral context. A neutral context provides all the features which are necessary to determine the extension of the (explicit or implicit) indexical expressions, and only these. This means that a neutral context contributes nothing to the illocutionary force of an utterance, and it contributes nothing to a non-literal understanding of utterances.

The claim will be, then, that any utterance of (1) within a neutral context establishes a speech act of the directive type, as opposed to, say, questions or assertions. This claim is equivalent to the position that it is part of the meaning of (1) that (1) can be used to perform a directive speech act, i.e. (1) expresses a corresponding speech act concept. However, the neutral context cannot distinguish between a command-like and an entreaty-like use of (1).

The theory of speech acts in general deals with speech act concepts and their relationships to sentences of natural language. A speech act concept results by application of a illocutionary type T to a propositional content P. A propositional content is either a proposition itself, or an open proposition (i.e. a function from possible worlds into pairs of a sequence of individuals and a truth value), or a propositional concept (i.e. a function from possible worlds into propositions), or, eventually, a predicate concept (i.e. a function from possible worlds into sets of individuals).

Each illocutionary type may be specified (or modified, or commented on) by explicit performatives, adverbials, and even higher clauses. Compare (1) with (3), and also (10) with (11) in their respective directive readings:

(10)   I urgently request you to shut the door.

(11)   I am afraid I have to ask you to shut the door.

In their directive readings, (3), (10) and (11) all express the same propositional content as (1), but, whereas (1) expresses an unspecified directive type, the others express a differently specified directive type. For the moment, I leave open the problem of how the semantic interpretation of specification takes place. Notice, however, that I don't start my analysis with the most specified explicit performatives like the so-called Performative Analysis does, but with the most unspecified grammatical moods.

I will assume that every act has a certain outcome defined by the concept of that act, and by the foregoing state of the world (and, possibly, by the counterfactual condition of what would be the case, if the act in question had not been performed). Every speech act has a certain outcome, too. As we will see, however, there are characteristic differences between the outcome of a non-linguistic act like closing a door (where the outcome is just another state of the world), and the outcome of a linguistic act like making a promise.

First, the outcome of a speech act depends on who the speaker and, in most cases, who the addressee is. (In general, acts where such a condition holds belong to an interaction.) Second, at least in most cases, a speech act brings about certain commitments on the part of the speaker, and also certain obligations on the part of the addressee, either directly, e.g. by a directive or an erotetic speech act, or indirectly concerning at least cases of non-understanding or not-shared presuppositions that require certain reactions. This means that the outcome of a speech act involves consequences for the future that cannot be described in causal terms.

Third, unlike a non-linguistic act like closing a door, in a speech act we may separate success from happiness, or, in other words, the purpose of the speech act may lie outside of it. To bring about that a door is closed we have to meet certain happiness conditions. If we do so, we will be successful at once, and the purpose of our endeavour will be satisfied. There may exist other purposes connected with the closing of a door – and normally they do exist – but they need not to be conventional, unless we use this act within a signal procedure, i.e. transform it into a part of an interaction.

Quite different is the situation with speech acts. To bring about the specific outcome of a speech act (whichever it might be) an utterance has also to meet certain happiness conditions – so far the same. If it does not, the outcome is undefined or empty, no speech act is performed at all – the utterance is void. (If the outcome in question has been already the case, then the utterance might be happy but abundant, except perhaps certain institutional speech acts like baptizing or sentencing. However, we are principally not able to shut a door already closed, unless we open it first.) Thus, if we have met the appropriate happiness conditions, our utterance has achieved an outcome; by means of the utterance we have, say, made a request or posed a question. However, we cannot say that we were already successful. The purpose of the request, which it conventionally has, is not satisfied by making that request but is settled with it, i.e. the purpose is settled with the outcome of the speech act.

The purpose of a directive speech act is satisfied if and only if the action denoted in the propositional content, which is an open proposition, has been performed by the respective addressee within a certain time interval after the time of utterance, regardless whether it was a request, or a command, or an entreaty. Only in case the purpose of a directive speech act has been satisfied I would say the speech act in question was in the end successful. Of course, we might introduce weaker notions of successfulness related to the understanding by the addressee, or to the overtaking of certain commitments resulting from the speech act. Nevertheless, these notions deal also with the future evolution of the interaction situation.

Notice that the satisfaction of the purpose of a request (i.e. its successfulness) is independent of whether the addressee performs intentionally the wanted action in order to comply with the request.

Notions like 'commitment' and 'obligation' are clearly pragmatic ones. We have to consider commitments as belonging to the interpretation speaker and addressee make, according to the cooperative nature of verbal interactions. Yet, I think, those interpretations have an appropriate semantic counterpart which is as follows.

We might identify the outcome of a speech act with the set of possible future courses of events (or courses of states of affairs) in which the purpose of the speech act in question will be satisfied, i.e. each speech act divides the total set of alternative future courses of events into two subsets. Speakers and addressees who are interested in cooperation with each other will try to react in such a way that they always remain within those courses of events which belong to the first set, i.e. in which the purpose of the foregoing speech act can be satisfied.

If we wish to account for grades of satisfaction and for strength of purpose (e.g. to distinguish more or less satisfying answers, or to distinguish between commands and entreaties as to their strength), we have to include a second part in the outcome, namely a valuation $\mathcal{V}$ of the elements in the set of possible courses of events. In the distinction of more or less satisfying answers we have to value at different courses of reactions which belong to the first set, whereas the distinction of commands and entreaties concerns the demarcation of the whole first set of courses of possible reactions from the second one. As one can see, the specification of a speech act type has mainly to do with the determination of the valuation function $\mathcal{V}$. Of course, $\mathcal{V}$ depends also on the context of use.

I am not entirely sure that the proposal I have made concerning the outcome of speech acts is adequate for all kinds of speech acts. The proposal seems to be reasonable in view of those speech acts that can open or reopen a discourse, or, more general, a sequence of interactions, like requests, questions, offers, reproaches etc. But what about speech acts like apologies, avowals, congratulations? Their purpose is to satisfy or compensate something already established in the past but not to call forth a certain future behavior. Therefore I will call them speech acts of the satisfactive kind. To account for them, either we have to introduce a rather weak notion of a future conformity with the fact that the speech act in question has satisfied something, or we have to give up the idea of an unified semantic treatment of the outcome of speech acts. It might be the case that satisfactive speech acts can be treated pragmatically only, according to their relationship with conditions brought about by the foregoing context of interaction. One argument in favor of such a treatment arises from the observation that there is no grammatical mood to express satisfactives but only particular verbs like 'apologize' and 'congratulate'. The same argument holds, as I shall show later on, with respect to the so-called conditional speech acts like warnings, advices etc. I also like to mention that we naturally regard an answer as something that satisfies a question without hereby claiming the existence of a separate answer speech act type. Answering is rather one of the functions of the assertion in discourse, and can of course be specified by using the verb 'answer'. Since in case of satisfactives, however, I do not have a very clear conception, I will leave the problem open.

I am now prepared to sum up the theoretical framework. The literal meaning of a sentence s of L is considered as a speech act concept $T(P)$ that defines what the outcome of a use of this sentence in a neutral context is. What semantics does is to study the properties of speech act concepts, and also, which speech act concept $T(P)$ is expressed by a certain sentence s of L in each of its readings relative to a neutral context of use c^0, and moreover, in case T has been specified by verbal means, which valuation $\mathcal{V}$ is expressed by this specification. Pragmatics studies which possible speech act concepts $T_1(P_1)$, $T_2(P_2)$, ..., together with their respective valuations $\mathcal{V}_1$, $\mathcal{V}_2$, ..., are expressed by a certain s relative to a non-neutral context of use c, given that s expresses $T(P)$ relative to a corresponding c^0, and that c is richer than c^0 and can be characterized by certain principles of conversation and certain knowledge, maxims, perceptions, and preferences of speaker and addressee.

4. THE SEMANTICS OF ILLOCUTIONARY TYPES

Two of the most interesting and fundamental illocutionary types are the directive and the erotetic. It has often been tried to deal with questions as a subspecies of directives. I have the strong feeling that such an approach must fail. Only superficially are questions and requests alike in that both call forth a reaction of the addressee. But an offer, a warning, a reproach do the same. The difference is easily seen if we look at the respective propositional contents. The appropriate propositional content for a directive is an open proposition where the predicate denotes an action concept that has to be realized by the addressee. An interrogative sentence, however, never mentions an action concept the addressee has to realize, but rather raises alternative propositions he has to decide between and one of which, or a weaker one at least, he has to assert.

As I have pointed out elsewhere, one can consider the propositional content of a *yes-no*-question and that of a disjunctive question to be a propositional concept. The addressee has to find out a proper value of the propositional concept, i.e. a proposition, relative to a given world, which in most cases will be the momentary world of the interaction situation. The propositional content of a *wh*-question, except *why*-questions, is a quite different one. I have proposed to consider it as a restricted predicate concept. The addressee has to find out a set of individuals which fall under the predicate, relative to a given world. From these differences of propositional content we can see both that the directive and the erotetic are quite different types, and that there are at least two significant subtypes of the erotetic.

I can here only indicate the basic idea of a theory of directives, to which I gave already some hints in the foregoing presentation. Be X a set of persons, T a set of time points, and W a family of possible worlds extended in time. All persons remain the same at all times in all worlds. The world $w \in W$ at t_i will also be noted short as w_i. Connected with each world $w \in W$ at each time $t_0 \in T$ is a set $F(w_0)$ of possible futures such that each element $f \in F(w_0)$ is a sequence of possible worlds ordered in time, and each two elements $f', f'' \in F(w_0)$ are for all past times $t \leqslant t_0$ identical.

Further on, be A a class of possible action concepts including 'doing nothing', and be $R(x, w, t)$ the repertoire of a person $x \in X$ in $w \in W$ at $t \in T$ such that it is a non-empty subset of A. If $a \in R(x, w, t)$, then the outcome of x performing a in w at t is defined by the outcome function $O(a, x, w, t)$. Let s be a simple imperative sentence of the form "V_a[imp]" (like 'shut the door'), where the verbal phrase V_a denotes an action concept $a \in A$. Its translation into the semantics language be $DIR(a(x_0))$, where 'DIR' stands for the directive type and x_0 marks the free place to be filled out by a name of the addressee $\beta \in X$. Let utt-s denote the action concept of uttering s.

We can, then, define the semantic interpretation of the imperative sentence s by the following characteristic outcome function:

(12)   If s is an imperative sentence of the form "V_a[imp]" and c^0 is a neutral context with respect to s at t_0 including at least a speaker α and an addressee β such that c^0 is a proper subset of w at t_0, then $O(utt\text{-}s, \alpha, c^0, t_0)$ is a set F' of sequences of possible worlds such that:

(i) F' is a proper subset of the set $F(w_0)$ of possible futures of w at t_0.

(ii)   For all elements $f \in F'$ the following holds: f includes a subsequence (w', w'') such that $w'' = O(a, \beta, w', t_i)$, where t_i is restricted to $t_0 < t_i < t_0 + \delta$ and δ depends on both the context of use c (which is richer than c^0) and the nature of a.

More generally, if we want to include imperative sentences with activity predicates and adverbials (like 'drive carefully') clause (ii) of (12) might be substituted as follows.

(13)   For all elements $f \in F'$, there exists a w_i in f such that $p(\beta)$ is true in w_i, where $t_o < t_i < t_0 + \delta$ and p is the open proposition expressed by s.

On the basis of the given explication we now can easily see why the first position discussed in Section 2 has failed, and how sentences (2) through (4) – in their respective descriptive readings – are related to sentence (1).

(14)   If (1) was uttered in a neutral context c^0 (speaker α, addressee β) at t_0, then

(i)   'order-to-shut-the-door' $(\beta)(\alpha)$ was true in c_0^0,

(ii)   the outcome of uttering (1) was such that 'be-obliged-to-shut-the-door' (β) was true in all elements of all

$f \in F'$ up to $t_i > t_0$ (when at t_i either β has shut the door, or the order has become inoperative),

(iii) the purpose of the performed speech act would be satisfied (i.e. the performed speech act would be successful) if and only if 'will-shut-the-door' (β) was true in c_0^0.

The last clause means that the outcome of uttering 'shut the door' in c^0 at t_0 is identical with the set of sequences of possible worlds that makes 'will-shut-the-door' (β) true in c^0 at t_0. However, from this we cannot conclude that sentences (1) and (2) have to have identical truth conditions. Rather, the truth conditions of sentence (2) are identical with the successfulness conditions of the speech act performed by an utterance of sentence (1) in a neutral context.

As to the erotetic type, I will give a definition for *yes-no*-questions only, related to the interaction situation. I take here the rather weak position that the purpose of a question will be satisfied if there exists some propositional act, regardless who performs it and which means are used, such that from its propositional content follows at least one of the alternative propositions raised by the question. No matter whether the latter is actually true. But in any case, the performer of the act will be committed to the truth of the expressed proposition according to the stress he lays upon it.

Let s be an interrogative sentence of the form 'S?' like 'does Paul shut the door?' Its translation into the semantics language be ERO $(\wedge \lambda p(p \in \{q, \sim q\} \& \vee p = 1))$ where 'ERO' stands for the erotetic type, q is the proposition expressed simpliciter by the corresponding declarative sentence S ('Paul shuts the door'), $\vee$ is the extensionality operator, '1' stands for the truth value True, $\wedge$ is the intensionality operator, and λ will be understood in the sense of the λ-calculus.

The semantic interpretation is defined by the following characteristic outcome function:

(15)   If s is an interrogative sentence of the form 'S?' and S expresses simpliciter the proposition q, and c^0 is a neutral context with respect to s at t_0 including at least a speaker α

such that c^0 is a proper subset of w at t_0, then $O(utt\text{-}s, \alpha, c^0, t_0)$ is a set F' of sequences of possible worlds such that

(i) F' is a proper subset of the set $F(w_0)$ of possible futures of w at t_0.

(ii) For all elements $f \in F'$ the following holds:

    (a) There exists a w_i in f and a person x such that x performs an act b in w at t_i, where $t_0 < t_i < t_o + \delta$ and δ depends on the context of use c (which is richer than c^0).

    (b) Part of the act concept b is the propositional content B, and B implies logically p such that $p \in \{q, \sim q\}$ (eventually, B together with true propositions in c implies logically p).

5. CONDITIONAL SPEECH ACTS

In conclusion I wish to say a few words about conditional speech acts like warnings, threats, advices, counsels, proposals, reproaches etc. One might guess that those speech acts belong to the directive type appealing to the fact that they can be realized by an utterance of an imperative sentence, e.g.

(16)   Watch the steps.

(17)   Turn left at the second corner.

To take such an utterance as a warning or an advice requires, however, to use contextual information in an essential way. And if we take the utterance as a warning or an advice, it cannot be considered to be of the directive type because the speaker doesn't expect the performance of the mentioned actions but merely a certain consideration by the addressee. The addressee can use the information given by the meaning-in-context of the uttered sentence to plan his further actions, i.e. he can rely on it in his practical inferences, that are, of course, also relative to his preferences.

Therefore, if we take (16), (17) as expressing a warning or an advice, we can make them more explicit by transforming them into a specific conditional, using the contextual information in the respective consequences, e.g.

(18)   If you don't watch the steps you will stumble.

(19)   If you turn left at the second corner you will get to the AC building.

But also in this case, we need contextual information to interpret (18), (19) as expressing a warning or an advice respectively. We have to assume that the addressee prefers not to stumble, or that he prefers to get to the AC building. Otherwise we would get rather an advice instead of a warning, and vice versa. Compare with

(20)   If you don't keep watching the steps you will see the beauty of the staircase.

(21)   If you turn left at the second corner you will run across the police patrol.

From this we can conclude that conditional speech acts like warnings advices etc. can only be analyzed on the pragmatic level, i.e. by referring to a non-neutral context of use which includes at least certain preferences of the addressee. As I suppose, this has to do with the fact that there exist in languages like English, German etc. no specific grammatical moods to express a warning or an advice. Instead of it, the imperative as well as the declarative mood can be used for both speech acts, provided one relies in a proper way on the context. Besides, there exist verbs like 'warn' and 'advise', which can, under certain circumstances, be used in explicit performatives:

(22)   I warn you of the steps.

(23)   I advise you to turn left at the second corner.

As I have pointed out, explicit performatives specify on the semantic level a certain speech act type, related to a proper propositional content. But (22), in this reading, expresses simpliciter no propositional content at all, and (23) doesn't express the propositional content of an advice, which has to be a conditional. Hence (22) and (23) must undergo the pragmatic interpretation to receive the full reading of a warning or an advice respectively.

Universität Düsseldorf
Allgemeine Sprachwissenschaft,
Universitätsstrasse 1,
D-4 Düsseldorf

ACKNOWLEDGEMENTS

I am grateful to David Kaplan, David Harrah, Helmut Schnelle, Renate Bartsch, Günter Todt, Rudi Keller and Ewald Lang for helpful discussions.

BIBLIOGRAPHY

Chellas, B.: 1971, 'Imperatives', *Theoria* **37**, 114–129.
Cresswell, M. J.: 1973, *Logics and Languages*, Methuen, London.
Davidson, D.: 1974, talk given at M.I.T., December, 1974.
Grice, H. P.: 1968, *Logic and Conversation*, Unpublished lectures.
Kasher, A.: 1974, 'Mood implicature – a logical way of doing generative pragmatics', *Theoretical Linguistics* **1**, 6–38.
Katz, J. J.: 1974. *Propositional Structure – A Study of the Contribution of Sentence Meaning to Speech Acts.*
Lewis, D.: 1969, *Convention.* Harvard University Press, Cambridge, Mass.
Lewis, D.: 1972, 'General Semantics', in Davidson and Harman (eds.), *Semantics of Natural Language.* D. Reidel, Dordrecht, pp. 380–397.
Searle, J. R.: 1969, *Speech Acts.* Cambridge University Press.
Stalnaker, R. C.: 1972, 'Pragmatics', in Davidson and Harman (eds.), *Semantics of Natural Language*, D. Reidel, Dordrecht; pp. 380–397.
Wunderlich, D.: 1974, 'Towards an Integrated Theory of Grammatical and Pragmatical Meaning', in Asa Kasher (ed.), *Language in Focus*, (Essays in Memory of Yehoshua Bar-Hillel), D. Reidel, Dordrecht, pp. 251–277.
Wunderlich, D.: 1975a, 'Behauptungen, Konditionale Sprechakte und praktische Schlüsse', to appear in *Studien zur Sprechakttheorie*, Suhrkamp, Frankfurt.
Wunderlich, D.: 1975b, 'Fragesätze und Fragen', to appear in *Studien zur Sprechakttheorie*, Suhrkamp, Frankfurt.
Wunderlich, D.: 1975c, 'Sprechakttheorie und Diskursanalyse', in Apel (ed.), *Sprachpragmatik und Philosophie.* Suhrkamp, Frankfurt, pp. 463–488.

VI

THE PROSPECTS OF
TRANSFORMATIONAL GRAMMAR

JOAN W. BRESNAN

TRANSFORMATIONS AND CATEGORIES IN SYNTAX

1. LOGICAL FORM AND GRAMMATICAL FORM: METHODOLOGICAL ISSUES

Many linguists are interested in the question, What is the relation between the grammatical form and the logical form of natural language sentences? Let us suppose that the grammatical form of sentences is given by a transformational grammar G, which generates each sentence with an associated structural description, consisting of the sequence of constituent structure trees in a transformational derivation of that sentence. The set of all such pairs of sentences and their structural descriptions can be thought of as the set of *grammatical forms*. Let us also suppose that the logical form of sentences is represented in an appropriate logic, or disambiguated interpreted language L, and think of the set of well-formed formulas of L as the set of *logical forms*. Suppose finally that we have a 'translation' mapping Φ from the set of grammatical forms to the set of logical forms. Then our question is, What are the properties of Φ? The basic issue that has divided transformational grammarians for nearly ten years is whether Φ is trivial. Linguists of the Generative Semantics school have held that it is, and those of the Interpretive Semantics school have held that it is not. Φ can be thought of as a system of rules of semantic interpretation.[1]

This way of posing the question permits us to make two small but essential points of terminological clarification. First, what most transformational linguists have meant by 'semantics' is the study of the properties of Φ or the syntax of L. This of course is quite a different thing from the truth-definitional semantics of the logic L itself, which is relegated to logicians. Second, what most logicians have meant by 'syntax' is a very different thing from the syntax of transformational grammar. Logical syntax consists of formation rules and proof-theoretic relations, or rules of inference, that are sometimes called

Butts and Hintikka (eds.), Basic Problems in Methodology and Linguistics, 261–282.
Copyright © 1977 by D. Reidel Publishing Company, Dordrecht-Holland. All Rights Reserved.

'transformation rules'. Grammatical syntax, by contrast, is concerned entirely with the formation of sentences: grammatical transformations are a kind of formation rule.

Since the answer to the question of the relation between logical form and grammatical form depends upon finding the appropriate logic and grammar, this is clearly an empirical issue.[2] Nevertheless, much of the well-publicized disagreement among linguists over this issue has been methodological.

Generative Semanticists have argued that it is methodologically preferable to *identify* deep structure (the initial tree in the structural description) with logical form, for then the mapping from logical form to surface structure (the final tree in the structural description) would be carried out entirely by a 'homogeneous' set of rules, in the words of Paul Postal, and, other things being equal between two theories,

the theoretically more impoverished [i.e. the one with fewer kinds of rules] is always more highly valued (Postal, 1972, p. 136).

This preference for a minimal theory is countered by the view that linguistic theory has the explanatory aim of determining the limits of natural grammatical form, by which human languages are distinguished from countless conceivable but unspeakable formal systems.

The fundamental problem of linguistic theory,

as Chomsky puts it,

is to account for the choice of a particular grammar, given the data available to the language learner. To account for this inductive leap, linguistic theory must try to characterize a fairly narrow class of grammars that are available to the language learner; it must, in other words, specify the notion 'human language' in a narrow and restrictive fashion (Chomsky, 1972, p. 67).

Given this aim, Chomsky argues,

If enrichment of theoretical apparatus and elaboration of conceptual structure will restrict the class of sets of derivations generated by admissible grammars, then it will be a step forward ... . One who takes the more 'conservative' stance, maintaining only that a grammar is a set of conditions on derivations, ... is saying virtually nothing (Chomsky, 1972, p. 68).

The Generative Semantics position has also been supported with a kind of reductionism which maintains that meaning and sound are in some sense the only 'observables' of language, hence that semantic and phonetic structures alone are empirically justifiable or 'natural', and

therefore that a transformational grammar should map natural semantic or 'logical' structures into surface structures without employing an
independently defined set of syntactic structures, such as Chomsky's
deep structures. Although linguists have rarely articulated this reductionist view explicitly, I think that something like it has been influential.

For example, Paul Kiparsky, who has emphasized the importance of
phonetics in phonology, has written,

Quite analogously, recent work in syntax is showing the importance of semantics in
syntax, and is leading to deep structure representations which are close to semantic
representations. It is only to be expected that progress in linguistics should consist of
reducing the abstract part of language, the part consisting of the various theoretical
constructs which must be set up to mediate between the concrete levels of phonetics and
meaning, the only aspects of language which can be directly observed (Kiparsky, 1973,
p. 52).

In the same vein Postal has written,

I believe that one should make every effort at this primitive stage of grammatical
understanding to describe the sentences of particular languages in what I shall call
natural grammatical terms. This means, among other things, that one should assume the
existence of no elements of structure, no levels of structure, and no kinds of representations whose existence is not absolutely necessary. I therefore assume, with work called
'generative semantics', that underlying grammatical structures are subparts of the logical
structures of sentences, *for it is known independently that these must exist* (Postal, 1974a,
p. xiv; emphasis added).

Here Postal appears to suggest that 'logical structures' have a kind of
epistemological priority or independent justification that syntactic
structures in the sense of Chomsky (1957, 1965) do not.

The denial of this reductionist position has come to be known as the
thesis of the autonomy of syntax, according to which logical form is not
given in advance of grammatical form: the syntactic categories and
rules of grammar – the 'formation rules' of natural language, if you
like – constitute an independent system which may be related to, but is
not necessarily reducible to, semantic notions. (For a recent discussion
of this thesis, see Chomsky, 1975.)

2. THE PROGRAM OF CATEGORY REDUCTION

The Generative Semantics view of the relation between logical form
and grammatical form quite naturally raised the problem of finding a

correspondence between the syntactic categories of grammar and the syntactic categories of logic. And in fact, the Generative Semantics program included an attempt to reduce the inventory of syntactic categories of natural language to the categories of a variant of the predicate calculus by extending the use of transformations. It is this program, and the alternative proposal advanced by Chomsky, that will be my main concern here.

The earliest work in transformational grammar emphasized the *structure-dependence* of grammatical transformations: unlike phrase-structure rules, or categorial rules, which specify categories in terms of their immediate constituents, transformations depend upon constituent structure and the *structural analyses* of phrases. Thus transformations have been a valuable tool for determining the structural analyses of sentences and the categorial classification of phrases. For example, a transformation like the English Passive, which preposes a Noun Phrase following a Verb, could be used as evidence that both the proper name *Mary* and the quantifier phrase *everything in sight* are Noun Phrases in the sentences *John kisses Mary* and *John kisses everything in sight*, because both phrases undergo the preposing: *Mary is kissed by John* and *Everything in sight is kissed by John* are grammatical. By the same token, the failure of *noisily* to undergo passivization in *John kisses noisily* (since **Noisily is kissed by John* is not grammatical) could be used as evidence that *noisily* is not a Noun Phrase in this sentence.

Now it often happens that categories that are distinguished by one transformation are not differentiated by another. For example, the Question Movement transformation of English applies to interrogative Noun Phrases and Adverb Phrases alike in examples such as *What woman does John kiss?* and *How noisily does John kiss?*, derived from *John kisses what woman?* and *John kisses how noisily?* These facts can be described by formulating the transformation disjunctively, so that Question Movement can prepose either a Noun Phrase or an Adverb Phrase. However, it was pointed out by Lakoff (1971) and McCawley (1972) that the use of disjunction in transformations indicated a loss of generalization, in that it permitted transformations in principle to apply to arbitrary sets of categories having nothing in common.

In the Generative Semantics program of category reduction, if two categories *A* and *B* behaved alike with respect to several transformations, this was often taken as syntactic evidence that A's are 'really' B's in underlying structure (or vice versa); the syntactic differences between A's and B's were then regarded as superficial irregularities. It was still necessary to account for these differences, and for this the 'exception feature' was used: A's differ from B's only in that the A-type lexical items are exceptionally marked [+rule i] or [-rule j] for each rule i or j that distinguishes them from B's. A lexical item positively marked with such an exception feature was said to 'trigger' the rule indexed by the feature. Thus McCawley:

... many category differences which had figured in previous analyses have turned out to hinge merely on whether certain lexical items do or do not 'trigger' certain transformations. ... The difference between nouns and verbs is that nouns but not verbs trigger a [special] transformation ... (McCawley, 1971, pp. 220–221).

In the same way quantifiers were treated as 'higher verbs' that triggered special 'lowering' transformations. (See Postal, 1974b, for a recent example.) Other category differences were similarly coded onto individual lexical items in the form of rule features, which triggered the applications of differentiating transformations. This program of category reduction has been reviewed by McCawley (1971, pp. 219ff.), who concludes that

the resulting inventory of categories matches in almost one-to-one fashion the categories of symbolic logic, the only discrepancy being that the category VP has no corresponding logical category. However, [as] Lakoff argued, there is in fact virtually no evidence for a syntactic category of VP ... .

These proposals were controversial; of the device of transformational exception features, Chomsky remarked that it was so powerful that it

could be used to establish, say, that all verbs are derived from underlying prepositions (Chomsky, 1970, p. 218, n. 21).

There were, I believe, two basic methodological flaws in this program. First, even if one accepts the goal of finding a direct correspondence between the categories of syntax and 'the categories of symbolic logic', this still leaves the theory of syntactic categories vastly underdetermined, as recent work in the logical analysis of natural languages shows. For example, Montague (1973) demonstrated that it is possible within a theory that maintains a strong relation between syntactic and semantic categories to treat quantified noun phrases and proper nouns in the grammatically traditional way – as belonging to the same category of Noun Phrase; he also showed that semantically opaque simple verb phrases (e.g. *seeks a unicorn*) can be treated in the grammatically traditional way – as simple VP's.[3] Recently, Cooper and Parsons (1974) have constructed transformational grammars – of both generative semantic and interpretive semantic types – which are provably equivalent to Montague's grammar. What, then, are the 'natural' syntactic categories of human languages? Should quantifiers be elements of NP's or higher V's? Should VP's be eliminated or retained? Since there is no unique logical form given in advance of grammatical form, we must look elsewhere for answers.

However – and this is a second methodological flaw – in the Generative Semantics framework we can no longer look to transformations for deciding evidence. For the extended use of exception features necessary in the Generative Semantics program undermined the fundamental property of transformations – their structure dependence. Exception features are clearly incompatible with a strong form of structure-dependency which requires that transformations depend *only* upon the structure of phrases, and not on their lexical content. But the use of exception features in practice also violated the weaker form of structure-dependency; for example, Lakoff's 'negative absolute exceptions' (Lakoff, 1970) were exception features designed to prevent structures that *satisfied* the structural conditions of transformations from being transformed by them. With this weakening of the structure-dependency of transformations, the evidential basis of syntactic theory was severely undercut. It is not surprising, from this perspective, that some practitioners of Generative Semantics began to disclaim transformational grammar altogether.

If we stopped and assessed the prospects of transformational grammar from this point of view, the picture would be gloomy. But my own view is optimistic, for there is an alternative line of research on the theory of syntactic categories which seems quite promising.

3. THE X̄ THEORY OF CATEGORIES

For reasons partly independent of the foregoing controversy, Chomsky (1970) proposed a new theory of syntactic categories, in which their similarities and differences could be represented systematically. This has become known from its notation as 'the $\bar{X}$ theory'.

An inadequacy of phrase-structure grammars had been pointed out by John Lyons (1968, pp. 330–332). Consider the phrase structure rules given in (1).

(1)   S → NP Aux VP

         VP → V NP

         VP → V

         NP → Det N

Lyons observes that

the rules given above fail to formalize the fact that there is an essential, language-independent, relationship between N and NP and between V and VP. As far as the formalization of phrase-structure grammars is concerned, it is a matter of 'accidental' coincidence that linguists will include in their grammars of different languages rules which always expand NP into a string of symbols containing N and rules which always expand VP into a string of symbols containing V. In other words, phrase-structure grammars fail to formalize the fact that NP and VP are not merely mnemonically-convenient symbols, but stand for sentence-constituents which are necessarily nominal and verbal, respectively, because they have N and V as an obligatory major constituent ... .

What is required, and what was assumed in traditional grammar, is *some way of relating sentence-constituents of the form XP to X* (where X is any major category: N, V, etc.). It would not only be perverse, but it should be theoretically impossible for any linguist to propose, for example, rules of the following form ... [emphasis added – JWB].

(2)   S → VP Aux NP

         NP → V VP

         NP → V

         VP → Det N

Chomsky's $\bar{X}$ theory provides a means for solving this problem. According to this theory, syntactic categories are decomposed into two kinds of universal syntactic primitives, *features* and *types*. If we assume two major features, N and V (for 'nominal' and 'verbal'), then the major lexical categories of English are specifiable as in (3). ((3) was given by Chomsky in lectures at the 1974 Summer Institute of the

Linguistic Society of America at the University of Massachusetts, Amherst.)

(3) $V \begin{bmatrix} +V \\ -N \end{bmatrix}$ verbs

$N \begin{bmatrix} -V \\ +N \end{bmatrix}$ nouns

$A \begin{bmatrix} +V \\ +N \end{bmatrix}$ adjectives

$P \begin{bmatrix} -V \\ -N \end{bmatrix}$ prepositions

Bear in mind that not all combinations of features need occur in every language and that there may be additional features.

In the $\bar{X}$ theory, the *type* of a category is represented by the number of bars above it; intuitively, the type of a category is the level or degree of structure it has. The lexical categories in (3) are of *type 0*. Categories of *type 1* consist of a lexical category and its complement(s). Illustrative phrase structure rules for English type 1 categories are given in (4).

(4) $\bar{V} \rightarrow V(NP)$ We are *nearing the meadow.*

$\bar{N} \rightarrow N(PP)$ *Nearness to the meadow* is the great virtue of our house.

$\bar{A} \rightarrow A(PP)$ The house was much *nearer to the meadow* after the tornado than before.

$\bar{P} \rightarrow P(NP)$ *Near the meadow*, we built a house.

The parentheses indicate that the complements are optional. The symbols 'NP' and 'PP' in (4) are equivalent to the categories $\bar{\bar{N}}$ and $\bar{\bar{P}}$ of *type 2* illustrated in (5).

(5) $\bar{\bar{V}} \rightarrow (Aux)\bar{V}$

$\bar{\bar{N}} \rightarrow (\bar{\bar{Q}})\bar{N}$

$\bar{\bar{A}} \rightarrow (\bar{\bar{Q}})\bar{A}$

$\bar{\bar{P}} \rightarrow (\bar{\bar{Q}})\bar{P}$

'Aux' includes the English auxiliary verbs; '$\bar{\bar{Q}}$' is the category of quantifiers, determiners and measure-phrases (Bresnan, 1973, 1976). Categories of types *3*, *4*, and *5* might be as on the left-hand sides of the phrase structure rules in (6).

$$(6) \qquad \bar{\bar{\bar{V}}} \rightarrow M\bar{V}$$

$$S \rightarrow \bar{\bar{N}}\bar{\bar{\bar{V}}}$$

$$\bar{S} \rightarrow (COMP)S$$

'M' in (6) would be the English modal or tense element; 'COMP' stands for complementizer, a kind of clause-particle. Like Aux, these are minor categories of null type; they are the heads of no categories.

Notice that the following generalizations hold of these English phrase structure rules (4)–(6). First, every category of type $i+1$ has as an obligatory major constituent a category of type i; the latter is called the *head* of the former. For example, the head of $\bar{V}$, a type 1 category, is V, a type 0 category. Second, if A is the head of B, A and B have the same syntactic features. For example, both V and $\bar{V}$ have the verbal feature matrix $\begin{bmatrix} +V \\ -N \end{bmatrix}$. Chomsky's Base Schema Hypothesis entails that these generalizations hold true of the phrase structure rules of all natural languages, and so captures the generalization noted by Lyons, that the head of a nominal category must be nominal, the head of a verbal category, verbal, etc. (Chomsky's Base Schema Hypothesis is a universal set of phrase structure rule schemata satisfying the first generalization and a convention for projecting the syntactic features of categories onto the categories of which they are the heads, which yields the second generalization; see Chomsky, 1970.)

A great advantage of the $\bar{X}$ theory of categories over the previous account is that it provides a natural way to characterize cross-categorial regularities without recourse either to disjunctions of categories or to exception features. To see how this can be done, consider (7), in which the major categories are represented by ordered pairs $\langle i, M \rangle$, where i is the type and M is the feature-matrix of the category.

(7)   *Categories*

$$V \quad \left\langle 0, \begin{bmatrix} +V \\ -N \end{bmatrix} \right\rangle \quad \bar{V} \quad \left\langle 1, \begin{bmatrix} +V \\ -N \end{bmatrix} \right\rangle \quad \bar{\bar{V}} \quad \left\langle 2, \begin{bmatrix} +V \\ -N \end{bmatrix} \right\rangle \quad \bar{\bar{\bar{V}}} \quad \left\langle 3, \begin{bmatrix} +V \\ -N \end{bmatrix} \right\rangle$$

$$N \quad \left\langle 0, \begin{bmatrix} -V \\ +N \end{bmatrix} \right\rangle \quad \bar{N} \quad \left\langle 1, \begin{bmatrix} -V \\ +N \end{bmatrix} \right\rangle \quad \bar{\bar{N}} \quad \left\langle 2, \begin{bmatrix} -V \\ +N \end{bmatrix} \right\rangle \quad S \quad \left\langle 4, \begin{bmatrix} +V \\ -N \end{bmatrix} \right\rangle$$

$$A \quad \left\langle 0, \begin{bmatrix} +V \\ +N \end{bmatrix} \right\rangle \quad \bar{A} \quad \left\langle 1, \begin{bmatrix} +V \\ +N \end{bmatrix} \right\rangle \quad \bar{\bar{A}} \quad \left\langle 2, \begin{bmatrix} +V \\ +N \end{bmatrix} \right\rangle \quad \bar{S} \quad \left\langle 5, \begin{bmatrix} +V \\ -N \end{bmatrix} \right\rangle$$

$$P \quad \left\langle 0, \begin{bmatrix} -V \\ -N \end{bmatrix} \right\rangle \quad \bar{P} \quad \left\langle 1, \begin{bmatrix} -V \\ -N \end{bmatrix} \right\rangle \quad \bar{\bar{P}} \quad \left\langle 2, \begin{bmatrix} -V \\ -N \end{bmatrix} \right\rangle$$

This representation allows us to characterize *classes* of categories in a natural way, by specifying a type i and a submatrix of syntactic features K (including the null matrix). That is what the 'X' predicates in (8) do: the number of bars over the X represents the type. An 'X' predicate with i bars above it and matrix K below it characterizes the set of all categories $\langle i, M \rangle$ for which K is a submatrix of M.

(8)   *Some 'X' Predicates and Sets of Categories Characterized*

$$\underset{\begin{bmatrix} +V \\ -N \end{bmatrix}}{X} \quad \{V\}$$

$$\underset{[+V]}{X} \quad \{V, A\}$$

$$\underset{[-N]}{X} \quad \{V, P\}$$

$$X \quad \{V, N, A, P\}$$

$$\underset{[+N]}{\bar{X}} \quad \{\bar{N}, \bar{A}\}$$

$$\underset{[-V]}{\bar{\bar{X}}} \quad \{\bar{\bar{N}}, \bar{\bar{P}}\}$$

$$\underset{\begin{bmatrix} -V \\ +N \end{bmatrix}}{\bar{\bar{X}}} \quad \{\bar{\bar{N}}\}$$

A transformation which applied only to Noun Phrases would employ the last predicate in (8); a transformation which applied to both Noun

Phrases and Prepositional Phrases would employ the next-to-last predicate. In this way cross-categorial transformations can be formulated without resorting either to disjunction or to exception features. Thus the $\bar{X}$ theory has a methodological advantage over the Generative Semantics account: it permits us to narrow the class of possible transformations without relinquishing their structure-dependency. (See Bresnan, 1976, for a more detailed development.)

4. EMPIRICAL ISSUES

So far, I have tried to show that methodological disagreements about the relation of grammatical form to logical form have motivated a major division in the field of transformational grammar, and I have contrasted the two divergent programs of research that resulted in the problem of constructing a linguistic theory of syntactic categories. One of the programs attempted to reduce grammatical categories to logical categories, and the other aimed to characterize grammatical categories independently. Let us now consider questions of evidence.

The choice of syntactic features in the $\bar{X}$ theory can be regarded as an empirical hypothesis about what are *syntactically natural classes of categories*. In the system I have described, a set of categories {P, NP, COMP} would be an 'unnatural class' because it could not be characterized by any 'X' predicate. If we discovered linguistically significant generalizations that applied to this class, that would falsify the system sketched above. On the other hand, if there are linguistically significant generalizations over the classes defined by the theory, this provides evidence that the sets are indeed *natural* classes of categories, and the theory is confirmed.[4]

In studying cross-categorial rules of English, I have found that certain sets of categories do indeed behave under transformation as though they formed natural classes, providing rather interesting evidence that chooses between the theories of categories that I have described. My results employ a generalized formulation of Chomsky's (1973) A-over-A principle. For justification and a more precise formulation of this principle than that given below, see Bresnan (1976); the research described here is drawn from Bresnan (1976).

Recall that the *A-over-A* principle is a general condition on the applicability of all transformations. A version of it is stated informally in (9):

(9)   *A-over-A Principle*

No transformation *T* can apply to remove a constituent *C* that satisfies a predicate *A* in *T*'s structural condition unless *C* maximally satisfies *A* under any proper analysis for *T*.

What this means is that if a transformation can apply in a given structure to two constituents *of the same kind,* one of which properly includes the other, the inclusive one must be chosen. The two constituents are of the same kind if they satisfy the same predicate *A* in *T*'s structural condition. Thus when we have a constituent of kind *A* 'over' another constituent of kind *A*, where the transformation *T* can apply to either, the 'higher' must be chosen.

To illustrate, the structure in (11) has two proper analyses with respect to the transformation in (10), as indicated.

(10) $Y - NP - Aux - V - (P) - NP - Z$
          1   2   3   4   5   6   7 →
          1   6   3   4   5 $\emptyset$ 7

(11)

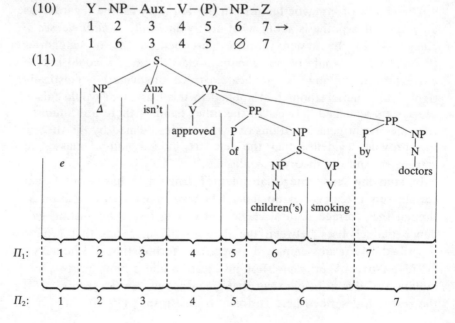

In the proper analysis Π_1, term 6 is a phrase of the kind NP; in Π_2, term 6 is also a phrase of the kind NP. Since the latter Noun Phrase properly includes the former Noun Phrase, the latter is the maximal phrase of the kind NP that satisfies the rule and by the A-over-A principle the structure must be transformed under the analysis Π_2. As a result, (12) but not (13) can be derived.

(12)   Children('s) smoking isn't approved of by doctors.

(13)   *Children aren't approved of smoking by doctors.

In contrast to the case just considered, applications of the Question Movement transformation in sentences like (14) and (15) are *not* prevented by this A-over-A principle, even though the interrogative phrases are of the kind NP and contained within larger Noun Phrases:

(14)   Who would you approve of my seeing?

(15)   Which books are you uncertain about giving to John?

To see why this is so, compare the structure (17) to the rule (16). (Question Movement is formulated more accurately later.)

(16)   wh $-$ W $-[_{NP}$wh $+$ Y$]-$ Z
           1   2   3   4 $\rightarrow$
           3   2 $\varnothing$ 4

In (16) and hereafter, 'wh' represents an interrogative morpheme.

(17)

Rule (16) can extract the NP *who* from the more inclusive NP *my seeing who* because the latter does not satisfy the structural condition of (16), which specifies that the NP which undergoes movement must begin with the interrogative morpheme *wh*.

It is because the *A*-over-*A* principle is sensitive in this way to the form of transformations that it can be used to investigate the $\bar{X}$ theory of categories.

Question Movement

The rule of Question Movement in English applies not only to Noun Phrases, but also to Adjective and Adverb Phrases that begin with an interrogative morpheme, as (18) shows.

(18)   *What book* did you read?
         How difficult is it?
         How quickly did you read it?

Since the phrases *what book, how difficult,* and *how quickly* are all type 2, Question' Movement can be formulated as in (19). The W_i are variables over arbitrary structures.

(19) $\text{wh} - W_1 - [_{\bar{X}} \text{wh} + W_2] - W_3$
         $\quad 1 \quad\quad 2 \quad\quad\quad 3 \quad\quad\quad 4 \rightarrow$
         $\quad 3 \quad\quad 2 \quad\quad\quad \varnothing \quad\quad\quad 4$

Now the *A*-over-*A* principle requires this rule (19) to apply to the maximal interrogative phrase of the $\bar{X}$-kind. So if we have a structure with two interrogative phrases of the $\bar{X}$-kind, one properly including the other, rule (19) can apply only to the more inclusive phrase. Such a structure is given in (20):

(20) $\text{wh} - \text{it is} - [_{\bar{N}} [_{\bar{A}} \text{how difficult}] \text{ a book}]$

Observe that both the $\bar{A}$-phrase *how difficult* and the $\bar{N}$-phrase *how difficult a book* are interrogative phrases of the $\bar{X}$ kind, but only the

Ñ-phrase is maximal. By the A-over-A principle, (21) but not (22) can be derived by Question Movement formulated as in (19).

(21)   How difficult a book is it?

(22)   *How difficult is it a book?

Similarly, when we reverse the situation and have an interrogative Ñ-phrase contained within an Ā-phrase, as in (23), only the Ā-phrase can be moved by the rule, because it is the maximal phrase of the X̃ kind that satisfies the rule.

(23)   wh – a ginkgo will grow – $[_{\bar{A}} [_{\tilde{N}}$ how many feet] tall]

(24)   How many feet tall will a ginkgo grow?

(25)   *How many feet will a ginkgo grow tall?[5]

If we had formulated Question Movement disjunctively, as in (26), the A-over-A principle would permit the ungrammatical examples (22) and (25) to be derived, because in (20) *how difficult* is the maximal interrogative phrase *of the Ā-kind,* and in (23) *how many feet* is the maximal interrogative phrase *of the Ñ-kind.*

(26) $wh - W_1 - [\begin{Bmatrix} \overline{\overline{N}} \\ \overline{\overline{A}} \\ \overline{\overline{Adv}} \end{Bmatrix} wh + W_2] - W_3$

   | 1 | 2 | | 3 | 4 | → |
   | 3 | 2 | | ∅ | 4 | |

What this shows is that the whole disjunction of simple predicates Ñ, Ā, $\overline{\overline{Adv}}$ in (26) must itself function like a simple predicate for the A-over-A principle to have the correct consequences. In other words, the class of categories described by the disjunction seems to behave as a unit: it appears indeed to be a *natural* class, as the X̄ theory predicts.

Comparative Deletion

Another cross-categorial transformation of English is the rule of Comparative Deletion. In (27) Comparative Deletion has applied to

delete a Noun Phrase from the *than*-clause; the deletion site is marked
'____'.

(27)   He uttered more homilies than I'd ever listened to ____ in
one sitting.

In (28) an Adjective Phrase has been deleted from the *as*-clause:

(28)   Try to be as dispassionate in writing your stories as you've
become ____ in conducting your affairs.

In (29) an Adverb Phrase has been deleted from the *as*-clause:

(29)   But they didn't word their proposal as skilfully as we
worded ours ____.

Syntactic analysis of these constructions shows that the deleted phrases
are of the form $\bar{\bar{Q}}$ *homilies*, $\bar{\bar{Q}}$ *dispassionate*, and $\bar{\bar{Q}}$ *skilfully*, where $\bar{\bar{Q}}$ is
a measure phrase of amount or degree (Bresnan, 1973, 1975, 1976).
Since all of these phrases are of type 2, Comparative Deletion can be
formulated as in (30).

(30) $[_{\bar{x}}[_{\bar{x}}\bar{Q}+W_1]-W_2][_{\bar{s}}W_3-[_{\bar{x}}\bar{\bar{Q}}+W_4]-W_5]$
$\quad\quad\quad 1 \quad\quad 2 \quad 3 \quad\quad\quad 4 \quad\quad\quad 5 \rightarrow$
$\quad\quad\quad 1 \quad\quad 2 \quad 3 \quad\quad\quad \varnothing \quad\quad 5, 4 \equiv 1$

The condition that term 4 be structurally identical to term 1 is required
for recoverability of deletions. Rule (30) says that a type 2 phrase
beginning with a $\bar{Q}$ is deleted under identity to another type 2 phrase
beginning with a $\bar{\bar{Q}}$; the deleted phrase is in the comparative clause
(the $\bar{S}$), while the phrase that it is identical to is included in a type 2
phrase to the left of the clause. The inclusion of term 1 in the $\bar{X}$ phrase
comprehending terms 1 and 2 does not have to be proper inclusion, for
the variable W_2, like all variables, may be null, in which case term 1 is
included in itself. This in fact is the case in examples (27)–(29). The
comparative construction in (27) has the structural analysis given in
(31) at the point where rule (30) applies, for example.

(31) $[_{\bar{N}}[_{\bar{\eth}}$ more] homilies] $[_{\bar{s}}$ than I'd ever listened to $[_{\bar{N}}\bar{\bar{Q}}$
homilies] in one sitting]

The leftmost $\bar{\bar{N}}$ in (31) satisfies both the $\bar{\bar{X}}$ predicate of term 1 in rule (30) and the $\bar{X}$ predicate of terms 1 and 2 in the rule.

By the A-over-A principle, (30) must delete the maximal phrase of the $\bar{\bar{X}}$-kind that satisfies the rule; this will be the largest compared phrase for which there is an identical phrase contained in the $\bar{X}$ phrase to the left of the comparative clause. To see how Comparative Deletion is affected by this principle, consider the structure given in (32).

(32) $[_{\bar{\bar{N}}} [_{\bar{A}} [_{\bar{\bar{Q}}} \text{as}] \text{ bad}]$ a poker-player$]$ $[_{\bar{S}}$ as he was $[_{\bar{\bar{N}}} [_{\bar{A}} \bar{\bar{Q}} \text{ bad}]$ a poker-player$]_{\bar{S}}]$

In (32), both the $\bar{\bar{N}}$-phrase $\bar{\bar{Q}}$ *bad a poker-player* and the $\bar{A}$-phrase $\bar{\bar{Q}}$ *bad* are identical to corresponding constituents to the left of the clause, namely, *as bad a poker-player* and *as bad*, respectively. Thus the rule could apply to both. But the $\bar{A}$-phrase is not the *maximal* phrase satisfying the rule, because it is properly contained in the $\bar{\bar{N}}$-phrase, which also satisfies the rule. Therefore the A-over-A principle allows (33) to be derived, but prevents (34).

(33)   John isn't as bad a poker-player as he was———.

(34)   *John isn't as bad a poker-player as he was ———a poker-player.

The comparative clause construction is notoriously complex, but the principle it illustrates here is rather simple: when Comparative Deletion can apply to two different categories of the $\bar{X}$ kind, one properly included in the other, it "chooses" the maximal one. We can in fact construct an example isomorphic to (34), in which the same $\bar{A}$-phrase *is* maximal, and its deletion is allowed:

(35)   John isn't as bad a poker-player as he was——— a bridge-player.

The reason that the deleted $\bar{A}$-phrase was maximal in (35) is that even though it was properly contained in an $\bar{\bar{N}}$-phrase $\bar{\bar{Q}}$ *bad a bridge-player* when the rule applied, the $\bar{\bar{N}}$-phrase failed to satisfy the identity condition in the rule: there is no phrase to the left of the clause to which it is identical. For justification of the details of this analysis, see Bresnan (1973, 1975, 1976).

It is not difficult to verify that a disjunctive formulation of Comparative Deletion would wrongly permit the derivation of ungrammatical examples like (34), so once again we see that the categories characterized by an 'X' predicate constitute a natural class.

Complex NP Shift

We turn finally to the rule of Complex NP Shift studied by Ross (1967, 3.1.1.3 and 4.3.2.3). As observed by Ross, this rule applies to both NP's and PP's, shifting them to the end of the verb phrase. Ross analyzes PP's as NP's, but we will formulate the rule as in (36).

$$(36) \quad W_1 - [_{VP} W_2 - \underset{[-V]}{\bar{\bar{X}}} - W_3] - W_4$$

$$\begin{array}{ccccc} 1 & 2 & 3 & 4 & 5 \rightarrow \\ 1 & 2 & \varnothing & 4 & 3\ 5 \end{array}$$

Recall that the predicate $\underset{[-V]}{\bar{\bar{X}}}$ designates $\{NP, PP\}$, so the effect of this rule is to postpose a NP or PP to the end of a verb phrase.

The A-over-A principle requires that the maximal NP or PP be shifted, and thus it correctly predicts that from structure (37), (38) but not (39) can be derived.

(37)   He [_{VP} considers [_{NP} many of [_{NP} my best friends]] stupid]

(38)   He considers stupid *many of my best friends.*

(39)   *He considers many of stupid *my best friends.*

Furthermore, when an NP is properly contained within a PP, only the PP can be shifted, if it is the maximal phrase of the $\underset{[-V]}{\bar{\bar{X}}}$ kind that satisfies the rule; so in (40) only the PP will be moved.

(40)   He [_{VP} talked [_{PP} to [_{NP} many of my best friends]] about my stupidity]

(41)   He talked about my stupidity *to many of my best friends.*

(42)   *He talked to about my stupidity *many of my best friends.*

Similarly, when a PP is properly contained within an NP, only the latter can be shifted by rule (36), as (43)–(45) show.

(43)   I [$_{VP}$ consider [$_{NP}$ arguing [$_{PP}$ with women who ride motor-cycles]] silly]

(44)   I consider silly *arguing with women who ride motorcycles.*

(45)   *I consider arguing silly *with women who ride motorcycles.*

(It must be borne in mind that there is a different rule, Extraposition of PP, that in various cases will detach a PP from an NP; see Ross, 1967.)

Again, if we had formulated Complex NP Shift disjunctively, sub-stituting $\left\{ \begin{array}{l} \text{NP} \\ \text{PP} \end{array} \right\}$ for the $\bar{\bar{X}}_{[-V]}$ predicate in (36), ungrammatical examples (42) and (44) would be derived, because the shifted phrases would have been the maximal phrases that satisfied the predicates NP and PP, respectively, in the rule.

In summary, application of the generalized *A*-over-*A* principle to various cross-categorial rules of English shows that the concept of "maximal phrase of the $\bar{X}$ kind" expresses linguistically significant generalizations; it appears that the $\bar{X}$ theory does indeed define syntactically *natural* classes of categories.

Now some of the facts discussed above had been described within the framework of category-reduction. Let us briefly compare the accounts.

For the Question Movement facts, the 'Left Branch Condition' was proposed (Ross, 1967, (4.181)):

(46)   *Left Branch Condition*

No NP which is the leftmost constituent of a larger NP can be reordered out of this NP by a transformational rule.

To sustain this formulation, Ross proposed that

how is analyzed as deriving from an underlying NP, and [adjectives] and [adverbs] are dominated by NP at the stage of derivations at which questions are formed (Ross, 1967, 4.3.2.1).

If Ross is correct that all these categories are to be derived from NP, the generalized A-over-A principle still accounts for the facts, making (46) unnecessary.

However, the Comparative Deletion facts had not hitherto been accounted for. The generalized A-over-A principle automatically explains them, given a formulation of Comparative Deletion like that I have proposed. The Left Branch Condition cannot be extended to these cases by generalizing it to deletions as well as reorderings, because it would then wrongly prevent the derivation of examples like (35) *John isn't as bad a poker-player as he was a bridge-player*, in which an underlying left branch adjective *bad* has been removed from a larger NP. For according to Ross, the adjective is an NP.

For the Complex NP Shift facts, Ross proposed a special constraint (Ross, 1967, (4.231)):

(47)   No NP may be moved to the right out of the environment
        $[_{NP} P\underline{\quad}]$.

Recall that Ross analyzes PP's as NP's that begin with a P – hence the label NP in (47). However, even if we regard PP's as NP's, (47) does not account for the ungrammaticality of cases like (45), where a PP is shifted from a larger NP that begins with a Verb rather than a P. (Nor does Ross's 'bounding' principle account for (45); see the discussion of Complex NP Shift in Bresnan (1976).)

The result of the comparison is that within the $\bar{X}$ theory, the generalized A-over-A principle provides a unified and general explanation for syntactic phenomena that were either described piecemeal or not accounted for in the category-reduction framework.

5. CONCLUSION

In conclusion, I think that the program of category reduction that I have described has proved to be misguided. This, however, does not mean that the goal of finding a close correspondence between grammatical form and logical form is also misguided – only that the logic appropriate for natural language must be a far more sophisticated invention than we have seen.

The aim of discovering in natural language that unique grammatical form which no logical syntax has adequately modelled does not seem to me to be misguided. This has long been a central goal of transformational grammar, and in view of recent advances in the field, I believe that the prospects for achieving it are good.[6]

Massachusetts Institute of Technology

NOTES

[1] I do not wish to suggest that semantic interpretation for natural language must involve a translation into logical form. See Cooper (1975), Fodor (1975), Hintikka (1974), Jackendoff (1976), and Miller and Johnson-Laird (1976) for discussion and alternatives.

[2] This point is underscored by Hintikka's argument that quantifiers in natural language are beyond the range of quantification theory; see Hintikka (1974).

[3] These points should not be obscured by Montague's use of the categorial notation in this paper. Cf. Montague (1970), Gabbay (1973), and especially Partee (1975).

[4] For confirmatory evidence for the $\bar{X}$ theory drawn from the domain of phonology, see Selkirk (1974).

[5] John Myhill has informed me that sentences like *How many feet is Mary tall?* are perfectly good in British English, although I believe they are not so in American English. Since examples like (25), in which the main verb before the adjective is not *be*, are ungrammatical in both British and American English, I will assume that the generalization holds, and that a different analysis must be provided for the exceptions with *be* in British English.

[6] I would like to thank Emmon Bach and Lars Hellan for discussions helpful in the preparation of this paper.

BIBLIOGRAPHY

Bresnan, J.: 1973, 'Syntax of the Comparative Clause Construction in English', *Linguistic Inquiry* **4**, 3.

Bresnan, J.: 1975, 'Comparative Deletion and Constraints on Transformations', *Linguistic Analysis* **1**, 1.

Bresnan, J.: 1976, 'On the Form and Functioning of Transformations', *Linguistic Inquiry* **7**, 1.

Chomsky, N.: 1957, *Syntactic Structures*, Mouton.

Chomsky, N.: 1965, *Aspects of the Theory of Syntax*, The MIT Press.

Chomsky, N.: 1970, 'Remarks on Nominalization', in Jacobs and Rosenbaum (eds.), *Readings in English Transformational Grammar*, Ginn and Co.

Chomsky, N.: 1972, 'Some Empirical Issues in the Theory of Transformational Grammar', in Peters (ed.), *Goals of Linguistic Theory*, Prentice-Hall, Inc.

Chomsky, N.: 1973, 'Conditions on Transformations', in Anderson and Kiparsky (eds.), *A Festschrift for Morris Halle*, Holt, Rinehart and Winston, Inc.

Chomsky, N.: 1975, 'Questions of Form and Interpretation', *Linguistic Analysis* **1**, 1.

Cooper, R. and Parsons, T.: 1974, 'Montague Grammar, Generative Semantics, and Interpretive Semantics', to appear in Partee (ed.), *Montague Grammar*, Academic Press.

Fodor, J.: 1975, *The Language of Thought*, Crowell.

Gabbay, D.: 1973, 'Representation of the Montague Semantics as a Form of the Suppes Semantics with Applications ...', in Hintikka *et al.* (eds.), *Approaches to Natural Language*, D. Reidel.

Hintikka, J.: 1974, 'Quantifiers vs Quantification Theory', *Linguistic Inquiry* **5**, 2.

Jackendoff, R.: 1976, 'Toward an Explanatory Semantic Representation', *Linguistic Inquiry* **7**, 1.

Kiparsky, P.: 1973, 'Phonological Representations', in Fujimura (ed.), *Three Dimensions of Linguistic Theory*, TEC Co. Ltd., Tokyo.

Lakoff, G.: 1970, *Irregularity in Syntax*, Holt, Rinehart, and Winston.

Lakoff, G.: 1971, 'On Generative Semantics', in Steinberg and Jakobovits (eds.), *Semantics: An Interdisciplinary Reader*, Cambridge University Press.

Lyons, J.: 1968, *Introduction to Theoretical Linguistics*, Cambridge University Press.

McCawley, J.: 1971, 'Where Do Noun Phrases Come From?' in Steinberg and Jakobovits (eds.), *Semantics: An Interdisciplinary Reader*, Cambridge University Press.

McCawley, J.: 1972, 'A Program for Logic', in D. Davidson and G. Harman (eds.), *Semantics of Natural Language*, D. Reidel.

Miller, G. and Johnson-Laird, P.: 1976, *Perception and Language*, Harvard University Press.

Montague, R.: 1970, 'English as a Formal Language', reprinted in Thomason (ed.), *Formal Philosophy: Selected Papers of Richard Montague*, Yale University Press, 1974.

Montague, R.: 1973, 'The Proper Treatment of Quantification in Ordinary English', in Hintikka *et al.* (eds.), *Approaches to Natural Language*, D. Reidel.; also reprinted in Thomason (ed.), op. cit.

Partee, B.: 1975, 'Montague Grammar and Transformational Grammar', *Linguistic Inquiry* **6**, 2.

Postal, P.: 1972, 'The Best Theory', in Peters (ed.), *Goals of Linguistic Theory*, Prentice-Hall, Inc.

Postal, P.: 1974a, *On Raising*, The MIT Press.

Postal, P.: 1974b, 'On Certain Ambiguities', *Linguistic Inquiry* **5**, 3.

Ross, J.: 1967, *Constraints on Variables in Syntax*, MIT doctoral dissertation; excerpted in Harman (ed.), *On Noam Chomsky*, Anchor.

Selkirk, E.: 1974, 'French Liaison and the $\bar{X}$ Notation', *Linguistic Inquiry* **5**, 4.

IN CONSEQUENCE OF SPEAKING*

When one gazes into his crystal ball and attempts to foresee the future course of transformational research into meaning and the use of language, the view is somewhat clouded by the scarcity of substantial past achievements and the lack of well-established approaches to research problems. There is, unfortunately, just no denying that the remarkable successes of transformational grammar in the area of syntax have not been paralleled by commensurate achievements in the realm of meaning. The regrettable dearth of inspiring research successes on which to model future efforts does not result from a lack of enterprise; for a dozen or more years now, a number of capable linguists have devoted considerable energy to supplementing transformational syntax with a semantic system. The literature on meaning is possibly more voluminous than that on syntax. All this toil has not yielded the solid results that were hoped for, however, and at a forward-looking moment such as the present symposium it may be in order to cast our view briefly backwards and inquire into the causes of this failure.

I wish to emphasize just one particular point. It is noteworthy, and no doubt understandable, that many linguists act as if their responsibilities in semantics were limited to those aspects of meaning which can be studied without going outside language, to studying meaning just as it shows up in semantic relations between sentences. This attitude is exemplified by the following representative illustration, excerpted from a recent article of Jackendoff (1976). At the outset he states, "In a linguistic theory, the semantic representation of a sentence is meant to be a formal characterization of the information conveyed by the sentence. This information can be thought of as a set of claims about various individuals, properties, events, and/or states of affairs, and about the relationships among them". (p. 89) This introduction naturally leads the reader to expect that Jackendoff's concerns

Butts and Hintikka (eds.), Basic Problems in Methodology and Linguistics, 283–297.

will include the way information is conveyed by means of making claims. Later, in discussing an array of sentences describing different sorts of motion of a thing from one place to another, he says "The semantic similarity among these seven sentences can be expressed by assigning them a common element in their semantic representations, a function GO(x, y, z). This function makes the claim that there has taken place an event consisting of the motion of x from y to z". (p. 93) At this point, Jackendoff is using semantic representations to model certain intralanguage likenesses of meaning. He hasn't gotten around to formally characterizing the information conveyed by the sentences, but it sounds like he may yet do so. Of course, GO(x, y, z) is an expression in an invented symbolic system and as such its meaning is no less in need of analysis than that of English sentences like 'The plane flew from New York to Toronto', which he wishes to explain with the aid of the symbolism. Eventually Jackendoff reveals that, "I take it that these functions [GO, CAUSE, LET, STAY, and BE: SP] are cognitive primitives of some sort, and that the way in which they make claims about the real world is more a problem in cognitive psychology than one in linguistics". (p. 120) We find out after all that he is not going to specify what meaning each semantic representation represents. Though he says they are to represent formally the claims made by sentences, he assigns to psychologists the problem of making them do so.

I don't mean by presenting this example to single out Jackendoff for criticism. The course he follows is one that most transformational linguists pursue of using symbols such as GO(x, y, z) to reflect a similarity of meaning between different sentences. Jackendoff also, in passages I have not quoted, uses this symbolization as an aid in codifying some purely formal rules of inference, which determine that a sentence having a certain form of semantic representation entails a sentence whose semantic representation is of a certain related form. Thus what is accomplished by the use of semantic representation is to reflect in an informal way certain likenesses of meaning between different natural language sentences and to express in a formal way certain deductive rules relating natural language sentences through the mediation of their semantic representations. These are very important concerns of semantics, but they do not exhaust the responsibilities of linguists in this area. They also are strictly internal to language and they leave the problem of explicitly specifying the claims a sentence makes pretty much where it was before we started.

Most transformational linguists, whether of the generative semantics or interpretive semantics persuasion, share the attitude that there is no reason for linguists to pursue meaning past the point where it leads out of language. Postal evinced this, for example, with his dictum (1970) that sameness of reference, as between a pronoun and its antecedent, is an important topic susceptible of study in linguistics but reference is a messy subject which linguists do well to avoid. The notion that we needn't look beyond language has drawn strength from a frequently voiced hope that it will one day be possible to devise a universal system capable of representing every meaning that can be expressed in any human language. That, of course, would allow us to solve problems of meaning in any natural language by solving them once and for all for this system and then assigning elements of the system as representations of the natural language's sentences. Realistically, however, linguists have at present neither a universal system for expressing meanings nor much of an idea what cognitive psychologists such as Miller and Johnson-Laird (1976) have found when they investigated some 'semantic primitives' of the kind that linguists have wished to make use of. The narrowness of approach which results from these deficiencies has produced some rather sterile debates. A better conception of the psychological bases of meaning and of how meaning allows language to represent the world may change our ideas about the way elements of a semantic system such as transformationalists assume are best used in semantic analysis.

It seems to me very important for linguists working on semantics to concern themselves with how meaning connects language to the world. This view is neither original nor idiosyncratic on my part; an increasing number of linguists hold it. Along with it goes the expectation that taking a broader view of the theory of meaning will help us break out of the impasse that has been reached in debates over a variety of current dilemmas.

One aspect of how meaning reaches out from language which linguists might well devote some study to is represented by the challenge to construct truth definitions for natural languages, which Davidson has advertised. Montague took up the challenge, with very enlightening and constructive results. And, in a heartening development, some linguists are now beginning to take it up as well. It is to be hoped that, as linguists investigate the intricacies of a truth definition for a natural language, they will keep in mind also another way in which meaning reaches out from language to the world, namely through speakers' intentions as revealed in their speech acts. For one thing, it is vitally important to keep in mind the crucial differences Grice has pointed out under the heading of conversational implicature between the constitutive rules of the language game we play and the strategies we employ in order to play it more effectively.

The idea that a sentence is true just in case it corresponds to reality is an appealing and, since Tarski, a formally very fruitful one. When one comes to assessing the material or, as linguists would say, observational adequacy of a proposed truth definition, however, one needs to realize that a great deal of the empirical bite comes, not from direct intuitions about whether a given sentence agrees with given facts, but rather from the role that a sentence's truth conditions play when one issues it in the performance of a speech act. It is, of course, obvious that when a person makes an assertion he commits himself to certain things being so, and that among those things is each necessary condition for the sentence which the speaker utters to be true. In ascertaining the truth conditions of a given declarative sentence, it is common practise to imagine conditions under which it would be appropriate to assert that sentence; indeed, when one is relying on intuition as a guide, it is difficult to think what other procedure one might follow. Widespread reliance on this simple procedure has, however, led to a general supposition that the truth conditions of many sentences are more stringent than I shall argue they actually are. The remainder of this paper is devoted to discussing certain other commitments a speaker makes in the performance of a speech act, besides to the truth of the sentence uttered, and the way those commitments affect the theory of meaning.

To begin with a concrete illustration, it would accord with many semanticists' practise, I believe, to say that the sentence

(1)   The Apollo XI astronauts failed to land on Mars

is false, since anyone who asserts it commits himself to its being so that the astronauts tried to land there or that they were expected to do so, and neither of these conditions is, as a matter of fact, fulfilled.[1] I do not dispute that the sentence would be an odd and even incorrect thing to say about the facts as they actually are; on that point I completely agree. But is it odd because the sentence is false? I think not, for reasons I will presently come to. In fact, I think the sentence is true, just as the following one obviously is.

(2)   The Apollo XI astronauts didn't land on Mars.

I shall argue that strictly speaking sentences (1) and (2) have identical truth conditions, and that the extra commitments a speaker makes by asserting the first one should be analyzed pragmatically. In short, I shall distinguish between truth conditions and what we may, following Grice, call conventional implicata.

What is a conventional implicatum? It is, first of all, something that is implicated. In his William James Lectures, Grice discussed a number of cases in which a listener is justified in inferring on the basis of a speaker's statement a proposition which does not logically follow from – i.e. which is not a necessary condition for the truth of – the sentence that was uttered. For example, in many contexts if I were to state

(3)   The source of the Nile is in either Ethiopia or Uganda

you would be justified in concluding that I do not know which alternative is true. This conclusion is not one which follows from sentence (3). It follows instead from my saying the sentence on the occasion when I did. In Grice's terms, my utterance implicates that I am ignorant of whether the Nile originates in Ethiopia or in Uganda, though it does not imply that proposition (the proposition is an implicatum of the sentence token I uttered). Implicature is an inherently pragmatic notion, bound up with illocutionary acts. Implication is a semantic or logical notion, relating sentences as abstract linguistic entities whether or not they be uttered.

Grice (1975) has sketched how this case of implicature, and many others, might be explained on the basis of some general assumptions about rational cooperation. Since such an explanation is available, there is no need to complicate our account of the meaning of sentence (3) in order to explain the inference. A vast range of implicatures are informally explicable like this one in terms of strategies for effective communication rather than conventions governing the literally correct usage of words and sentences; Grice calls them *conversational* implicatures.

But some implicatures, including the one exemplified by sentence (1), cannot be explained on conversational grounds. The conclusion that the astronauts tried, or were expected, to land on Mars is licensed by an utterance of (1) because of a convention about how English speakers are to use the expression 'failed to'. To state

(4)   The Apollo XI astronauts failed to land on Mars, though they neither tried nor were expected to land there

would be absurd; anyone who said it would be seen as contradicting himself. The implicature of sentence (1) is uncancellable because it is part of what the sentence means. As such it is appropriately recorded in the lexical extry for the verb 'fail (to)'. Such *conventional* implicatures contrast with conversational ones, which are cancellable. Compare example (5) with its parallel, (4).

(5)   The source of the Nile is in either Ethiopia or Uganda, and I know which (though I am not going to tell you).

When one controverts a conversational implicature there is more a feeling of uncooperativeness than of inconsistency, whereas the latter predominates in the case of a conventional implicature.

So if the proposition that the astronauts tried or were expected to land on Mars is implicated by utterances of sentence (1), it must be a conventional and not a conversational implicatum. But what are the reasons for thinking it is an implicatum at all, rather than a necessary condition for the truth of (1)? When one asserts a sentence, he commits himself not only to each necessary condition for the sentence's truth but also to every one of its conventional implicata. The two classes of commitments, however, are not equally weighty; the commitment to truth is primary, that to the conventional implicata subsidiary. This disparity can be brought out by considering differences in the kinds of criticism that are appropriate to breaches of the two sorts of commitment. If the commitment to truth of the sentence is satisfied but some conventional implicatum is false – as is, in fact, the case with example (1) – then the speaker gets 'partial credit' for speaking truthfully. He is liable to mild criticism such as: 'Well, they certainly didn't land there, but then they never were intended to'. On the other hand, if the commitment to truth is breached, then the issue scarcely arises of whether a falsehood has been conventionally implicated as well. A person who asserted

(6)   The Apollo XI astronauts failed to land on the Moon

would get no partial credit for the correctness of what he conventionally implicated. One would hardly say to him: 'You're right that they tried to land there, but as a matter of fact they actually succeeded'. A much stronger rebuke would be forthcoming instead, as this speaker would have blundered about as badly as one who asserted

(7)   Dag Hammarskjöld failed to die in a plane crash.

Implicating a falsehood while speaking falsely adds little or no seriousness to the offense, because asserting a false sentence is already such a gross error. If we find that a speaker has failed in his primary responsibility – to speak only the truth – we do not fret about whether he has lived up to lesser ones, such as to conventionally implicate only the truth.

The difference between truth conditions and conventional implicata can be brought out with careful attention to the distinction between misstatements that allow 'partial credit' and those that do not. Significant as this phenomenological difference is, though, it is not always decisive as a criterion for sorting out truth conditions from conventional implicata. Fortunately, it is paralleled by a difference in the roles these two aspects of meaning play in the rules that specify the meaning of syntactically complex expressions as a function of the meanings of their parts. This difference too is most easily brought out with the aid of examples.

Consider how you might answer the following question.

(8)   Did the Apollo XI astronauts fail to land on Mars?

It would be quite appropriate to respond: 'Well yes, but there was never any reason to think they would land there'. On the other hand, it would be quite out of place to respond 'No, in fact they never planned to land there'. This observation is easy to understand, on the plausible assumption that the interrogative sentence (8) asks for the truth value of sentence (1).[2] It simply reflects the fact that (1) is actually true, not false, because its truth conditions are identical to those of sentence (2).

Or consider another illustration. With no expectation that it will be effective, I give my son a warning in the following words.

(9)   If you fail to pick up your toys, I am going to spank you.

An hour later I return to find his room still strewn with playthings, and I proceed to spank him. Does my son have grounds to protest that I did not give him fair warning, that my statement gave him no reason to expect a spanking, since he never tried to pick up his toys nor did I really expect he would do it? It seems clear that he hasn't, that simply not picking up his toys is sufficient to satisfy the antecedent of the conditional sentence (9).

These observations show that truth conditions and conventional implicata of a sentence figure differently in the calculation of the meaning of more complex expressions constructed out of the sentence. The dichotomy is further bolstered by another fact about the interrogative and conditional sentences. In asking the question (8), a speaker commits himself to the proposition that the astronauts tried or were expected to land on Mars, which I have identified as a conventional implicatum of the corresponding declarative sentence, (1). This fact would be quite mysterious if the proposition were not a conventional implicatum but rather a truth condition of the declarative. Since, however, it is possible to incidentally commit oneself to the truth of certain propositions while doing the primary business of attempting to induce one's audience to supply certain information, this fact simply reflects the rule that an interrogative sentence conventionally implicates everything that the transformationally corresponding declarative does. Moreover, this rule is part of the reason why an unadorned 'Yes' will not do as an answer to the question (8). In asking his question, the speaker also puts forward a thesis to the effect that the astronauts made a certain attempt, or whatever; in an unassertive way he drops that thesis into the pool of mutually agreed upon propositions. Unless the respondent indicates his disagreement, for example by denying the offending proposition, he will be taken by the questioner as accepting the implicata of the question. This is the way the language game is played.

Corresponding remarks apply to the conditional sentence. As is well known, necessary conditions for the truth of the antecedent need not be necessary conditions for the truth of a conditional sentence. For instance,

(10)   Kissinger forced the CIA to spy on Nixon

can be true only if the CIA spied on Nixon. But

(11)   If Kissinger forced the CIA to spy on Nixon, he wielded
         more power in the government than the public yet realizes

may be true even if the CIA didn't spy on Nixon. In this light, consider the fact that an assertion of

(12)   If the Apollo XI astronauts failed to land on Mars, there
         won't be a manned landing there before the 21st century

commits the speaker to the astronauts having tried or been expected to land on Mars, just as much as an assertion of sentence (1), its antecedent, does. How odd this would be if the proposition in question were only a truth condition for sentence (1). Actually, of course, it is a conventional implicatum, and the fact we have noted about sentence (12) simply reflects the general rule that an indicative conditional sentence conventionally implicates everything its antecedent does.

For this reason, my warning (9) to my son commits me to the proposition that he will try, or will be expected, to pick up his toys. In the situation described, this is false because I didn't expect him to do it and, as things turned out, he didn't even try. But this secondary error of conventionally implicating a falsehood does not completely undermine my speech act. I still deserve partial credit, for saying something that is strictly speaking true. So my son is adequately warned despite my error.

There are many categories of words whose meaning contributes to the conventional implicata of sentences containing them. To cite just a couple of more examples, the verb 'manage (to)' and the adverb 'even' contribute only conventional implicata and are vacuous as far as truth conditions go. Both of the following questions conventionally implicate falsehoods, yet the answer to each is 'Yes' (with a qualification rejecting the implicature) rather than a qualified 'No', indicating that in each case the corresponding declarative sentence is true.

(13) (a)   Did Nixon manage to resign from the Presidency?

(b)   Does even Carter have a chance to beat Ford?

Other words contribute a mixture of truth conditions and conventional implicata. For example, the adverb 'almost' gives rise to a counterfactive conventional implicature, as well as affecting the truth conditions of sentences containing it. The question

(14)   Did Franco live almost to the age of eighty?

conventionally implicates a falsehood, but the correct answer is something like: 'Yes, in fact he lived to be eighty-three'. It would hardly do to respond: 'No, in fact he lived to be eighty-three'.

Not only lexical items but also grammatical constructions contribute conventional implicata to sentences, as well as influencing their truth conditions. The non-restrictive relative clause construction, for instance, creates a conventional implicature that the proposition expressed by the relative clause is true. Whoever asks the question

(15)   Did Agnew, who is Vice-President of the U.S., write a novel?

commits himself to the falsehood that Agnew is Vice-President of the U.S. To take another example, the cleft sentence construction (as exemplified by 'It was Brutus who killed Caesar') gives rise to a conventional implicature that some individual satisfies the function expressed by the clefted sentence. To assert either

(16)   It wasn't Ford who pardoned Agnew

or

(17)   If it wasn't Ford who pardoned Agnew, then Ford must have pardoned someone else

would commit the speaker to the proposition that someone pardoned Agnew.

This is only a small sample of the wide range of cases for which conventional implicata can be sorted out from truth conditions. When giving an explicit description of what expressions of a natural language mean, one should take account of the difference between these two aspects of meaning.

Returning to the point at hand, the methodology of generative grammar, of seeking fully explicit descriptions of languages, is helpful in further clarifying this distinction. Let us examine how the formal roles of truth-determining aspects of meaning and implicature-determining aspects differ, in the recursive rules that assign complete meanings to expressions of English. This is investigated in papers by Lauri Karttunen and myself; the basic analysis is presented in (1975), it is applied to solve another problem in (1976), and discussed in detail in (in preparation). In exploring formal rules for assigning conventional implicata to sentences, we have used Montague's framework for describing languages, not out of a conviction that it is inherently superior to transformational grammar but because at present it alone provides for an explicit and sophisticated model-theoretic interpretation of the language being described. I will briefly give you a rough idea of what our results are in these papers.

As in Montague (1973), we obtain an interpretation for English, or more precisely for that fragment of it which we treat, by translating to a model-theoretically interpreted auxiliary language. This convenient intermediate step is, as Montague (1970) showed, eliminable in a principled way. We assign each lexical item as its translation a pair of expressions of the auxiliary language (called Intensional Logic), not just a single expression as Montague did. The first one we call the *extension expression* since it is intended to denote under the interpretation of Intensional Logic the extension which the English expression has. The second component of the translation is called the *implicature expression* as, loosely speaking, it is intended to denote what the English expression conventionally implicates. To cite a specific example, the lexical item 'fail to' translates to a pair of expressions of Intensional Logic with the following interpretations. The extension expression denotes a function which maps a property to the set of things that do not have it (e.g. the property of landing on Mars is mapped to the set of things that haven't landed there). The implicature expression denotes a function which maps a property to the set of individuals that try to or are expected to have that property.

Non-lexical expressions of English are generated and assigned translations by a system of recursive rules. These have a form illustrated by the following one.

(18)   If α is a phrase of category IV//IV (for example, 'fail to') and β is a phrase of category IV (for example, 'land on the Moon'), then $\alpha\beta$ is a phrase of category IV. If, furthermore, α translates to $\langle\alpha^e; \alpha^i\rangle$ and β translates to $\langle\beta^e; \beta^i\rangle$, then $\alpha\beta$ translates to $\langle\alpha^e(^\wedge\beta^e); \hat{x}[\alpha^i(^\wedge\beta^e)(x)\wedge\alpha^h(^\wedge\beta^i)(x)]\rangle$, where α^h is the value of h on $\langle\alpha^e; \alpha^i\rangle$.[3]

Rule (18) constructs the verb phrase 'fail to land on the Moon', for example, by concatenating the infinitive-taking verb 'fail to' with the phrase 'land on the Moon'. This rule utilizes a very simple syntactic operation; others make use of complicated syntactic transformations rather than straightforward concatenation. The second part of rule (18) shows how the translation of the derived phrase is composed out of the translations of its parts. Given the intended interpretation of Intensional Logic, this shows how the meaning of the derived phrase of English is composed out of the meanings of its parts.

What I want to point out about the translation assigned by rule (18) is this. The extension expression of the derived phrase is constructed just from the extension expressions of the parts. Since the extension expression of a sentence denotes its truth value, the first expression in the translation of any phrase of English signifies the truth-determining aspect of that phrase's meaning.

In contrast, the implicature expression which rule (18) assigns to the derived phrase depends on both the extension expression and the implicature expressions of the phrase's parts. The specific conventional implicature contributed to a sentence by a verb, for example 'fail to', is a function of the property associated with the particular infinitival phrase (e.g. 'land on Mars' vs 'land on the Moon') constructed with the verb in that sentence. (The circumflex prefixed in $^\wedge\beta^e$ forms an expression of Intensional Logic denoting the property associated with β^e, the property that assigns to each possible world the set denoted by β^e relative to that possible world.) The total conventional implicata arising from the verb phrase are those contributed by the verb plus those inherited from its complement (for technical reasons, these implicata can be conjoined only with the aid of a variable, x). The reason we introduce the function h is that not all verbs are like 'fail to' in letting the implicata of their complement shine through. The verb 'hope to', for example, transmutes its complement's implicata into beliefs (roughly speaking). Whereas 'John failed to survive' conventionally implicates that John was in mortal danger, 'John hopes to survive' implicates the weaker proposition that John believes his life is in danger. We use the heritance function h to associate with each verb the mapping appropriate to its meaning which will convert the implicata of its complement into the form in which they are inherited by the derived verb phrase.

Rule (18) illustrates the usual case in English. When a rule combines two phrases one of which semantically modifies the other, the implicata of the derived phrase are the sum of those inherited from the modified phrase, under a transformation determined by the modifying phrase, with the implicata generated by the action of the modifier on the intension of the modified phrase. The formal difference between this recursion and the simpler one discussed two paragraphs above, which defines the extension of derived phrases, provides a very useful test for diagnosing what is a truth condition and what a conventional implicatum of an English sentence. If one is willing to undertake the construction of an explicit description of what the sentences of a language mean, he need not rely solely on intuitions about misstatements that allow 'partial credit' versus those that do not as a means of distinguishing truth conditions from conventional implicata.

I want to stress that, though I have made use of an auxiliary invented language in this sketch, the meaning of the expressions of Intensional Logic used in translating English has been thoroughly worked out. The form $A(B)$ signifies functional application, $\hat{x}A$ signifies λ-abstraction on the variable x, $^\wedge A$ signifies functional abstraction on an implicit variable over possible worlds, and so on.

Thus when combined with an interpretation of the non-logical constants appearing in the translations assigned to English sentences, our grammar does give a formal account of those sentences' truth conditions and conventional implicata. The grammar together with an interpretation of Intensional Logic gives the right sort of thing to feed into a speech-act theory, not uninterpreted formulas of an invented language but rather a truth definition and, moreover, a definition of correctness of conventional implicature. The description shows, at least in part, how the English language is connected to the world. Our account is not given in terms of what meaning relations hold within English. It does not say, for instance, that sentence (1) conventionally implicates certain English sentences. Indeed, if one had not looked beyond English-internal meaning relations, the notion of conventional implicature might well not have come to light. Instead it tells what propositions (which are not symbolic expressions) are conventionally implicated by a sentence, and also what propositions are necessary conditions for the sentence's truth.

I see no reason not to try to associate such semantic descriptions with transformational grammars. Indeed, Montague's framework has many, and striking, syntactic shortcomings when it comes to generating natural languages. Karttunen and I used it only because it has not yet been shown how to model-theoretically interpret transformational structures in the way that is possible with Montague grammars. Working out how to do this for transformational grammars seems to me a very promising direction for research in view of the syntactic virtues of transformational grammar and the semantic virtues of model theory.

University of Texas at Austin
Institute for Advanced Study, Princeton

NOTES

* In preparing this paper, I have benefitted greatly from conversations with Lauri Karttunen. He should not be taken, however, as agreeing with everything I say here. I am indebted for support of this research to a grant from the Sloan Foundation to the Institute for Advanced Study, and indebted also to the Research Workshop on Alternative Theories of Syntax and Semantics conducted by the Mathematical Social Science Board at the University of California, Berkeley, in the summer of 1975.
[1] It has come to my attention that for some English speakers 'failed to' is a mere stylistic variant of 'didn't'. Such speakers will not find it odd to assert sentence (1). I suggest that they try substituting examples with the verb 'neglected to'. For instance, 'The Apollo XI astronauts neglected to bring back any meteorites' is true given that, as a result of negligence, the astronauts didn't bring back any meteorites. But it is an incorrect thing to assert because the astronauts were not under any obligation to bring back meteorites.

[2] There may be another way of taking the question. Someone might, perhaps in a picky frame of mind, take it as asking whether it would be appropriate to assert 'The Apollo XI astronauts failed to land on Mars'. This reading of the question should be disregarded.

[3] In Karttunen and Peters (1975, 1976), the translations are triples of logical expressions because the value of h on $\langle \alpha^e; \alpha^i \rangle$ is explicitly carried along in the recursion.

BIBLIOGRAPHY

Grice, H. P.: 1975, 'Logic and Conversation', in D. Davidson and G. Harman (eds.), *The Logic of Grammar*, Dickenson Publishing Company, Encino, California, 1975, pp. 64–75.

Jackendoff, R.: 1976, 'Toward an Explanatory Semantic Representation', *Linguistic Inquiry* **7,** 89–150.

Karttunen, L. and Peters, S.: 1975, 'Conventional Implicature in Montague Grammar', in C. Cogen *et al.* (eds.), *Proceedings of the First Annual Meeting of the Berkeley Linguistics Society*, Berkeley Linguistics Society, Berkeley, 1975, pp. 266–278.

Karttunen, L. and Peters, S.: 1976, 'What Indirect Questions Conventionally Implicate', in S. Mufwene *et al.* (eds.), *Papers from the Twelfth Regional Meeting of the Chicago Linguistics Society*, Chicago Linguistics Society, Chicago, 1976, pp. 351–368.

Karttunen, L. and Peters, S.: in preparation, 'Conventional Implicature'.

Miller, G. A. and Johnson-Laird, P.: 1976, *Perception and Language*, Harvard University Press, Cambridge, Massachusetts.

Montague, R.: 1970, 'Universal Grammar', *Theoria* **36,** 373–398.

Montague, R.: 1973, 'The Proper Treatment of Quantification in Ordinary English', in K. J. J. Hintikka *et al.* (eds.), *Approaches to Natural Language* (Synthese Library), D. Reidel Publishing Company, Dordrecht and Boston, 1973, pp. 221–242.

Postal, P. M.: 1970, 'On Coreferential Complement Subject Deletion', *Linguistic Inquiry* **1,** 439–500.

JEAN-ROGER VERGNAUD

FORMAL PROPERTIES OF PHONOLOGICAL RULES*

1. INTRODUCTION

One of the salient features of generative phonology has been the emphasis put on formal questions. The study of abstract properties of grammars has been the distinctive concern of work carried out in the field and most of the research has been aimed at discovering formal universals. The fundamental methodological assumption has been that the investigation of formal properties of grammars would eventually lead to significant discoveries. At the same time, however, generative linguists have been aware that phonological theory, as a part of a general theory of language, should concern itself with substantive universals as well. In particular, it was apparent that the evaluation measure would have to incorporate an elaborate system of substantive constraints. For example, in *The Sound Pattern of English* (SPE),[1] the basic theoretical work of generative phonology, Chomsky and Halle remark that the formal evaluation procedures and the associated notational devices defined in the early chapters of their book give wrong results in many instances and that they must be supplemented by a set of conventions which takes into account the intrinsic content of phonological features and of phonological rules.[2] Of course, there is no contradiction in undertaking investigations of both types of universals, formal and substantive. The open, and empirical, question, however, is how greatly substantive conditions limit the class of grammars available to the language-learner: if *very* greatly, formal properties have little interest. In this paper I would like to provide a partial answer to this question, and to argue that the fundamental methodological assumption of generative phonology, as stated at the beginning of this introduction, is correct. I will suggest that phonological theory contains a rich core of formal constraints. In particular, I will present some striking results from Halle *et al.* (1975).

Butts and Hintikka (eds.), *Basic Problems in Methodology and Linguistics*, 299–318.
Copyright © 1977 by D. Reidel Publishing Company, Dordrecht-Holland. All Rights Reserved.

2. PHONOLOGICAL REPRESENTATIONS

The phonological component can be regarded as an input-output device which operates on a string of formatives, provided with a structural analysis by the syntactic component, and assigns to this string a phonetic representation. Formatives are of two types: grammatical and lexical. Among the grammatical are the junctural elements such as the word boundary '#', introduced into the surface structure by rules which are partly universal, partly language-specific.[3] Other grammatical formatives are the tense markers, such as *Present, Past,* and the number markers, such as *Plural,* etc. Each grammatical formative g is represented by a single symbol s_g. Each lexical formative l has a phonological representation P_l. A subpart of the phonological component eliminates all grammatical formatives (s_g's) except junctures in favor of phonological representations (P_g's). Thus, at a certain stage, we have a representation in terms of P_i's and junctures alone (with the derived constituent structure of the string indicated). This representation is called the level of *systematic phonemics.* The phonetic representation that constitutes the output of the phonological component is called the level of *systematic phonetics.* This level is essentially a representation in terms of a universal phonetic system.

Theories of phonology may differ in the degree of complexity of the notion 'phonological representation'. In the standard theory (cf. SPE), a formative at the level of systematic phonemics is represented in a systematic orthography as a string of symbols, each of which is assigned to certain categories (consonantal, voiced, nasal, etc.). For example, the phonological representation of the lexical formative *in* is a string of two symbols, the first of which belongs to the categories vocalic, front, high, etc., and the second of which belongs to the categories consonantal, nasal, coronal, etc. Each symbol can, in fact, be regarded as an abbreviation for the set of categories to which it belongs, and each P_i can thus be represented as a *classificatory matrix* in which the columns stand for what phonologists have called 'segments' and the rows, for categories; the entry in the ith row and the jth column indicates whether the jth segment belongs to the ith category. These categories are called (classificatory) 'distinctive features'. For example, the classificatory phonological matrix corresponding to *in* includes

	i	n
consonantal	−	+
vocalic	+	−
nasal	−	+
voiced	+	+

where '+' means 'yes' and '−', 'no'.

Note that, in the standard theory, the symbols of the universal phonetic system are specified in terms of a set of phonetic features which coincides with the set of phonological features; hence, the output of the phonological component can also be regarded as a matrix in which columns represent phones and rows, phonetic features of the universal system. The entry in the ith row and jth column indicates whether the jth phone of the generated utterance possesses the ith feature, or the degree to which it possesses this feature (in the case of such features as stress).

We may think of the standard theory, as described above, as the 'minimal theory'. It is argued in Halle *et al.* (1975), henceforth FP, that this minimal theory is inadequate and that a richer notion of phonological representation is needed. Suppose (1) that such features as 'long' are suprasegmental features, and (2) that certain discontinuous sequences of substrings are analyzed as elements of certain complex categories (cf. the sequence of vowels in tone-assignment). Then each phonological string (in the sense of SPE) will be provided with a phonological structural analysis that has the following characteristics:

> (i)   the features ± long, for example, can be predicated of substrings containing more than one segment
>
> (ii)   the phonological string corresponding to a word[4] is paired with a set of disjoint tree-like structures, which are hierarchical representations in terms of complex categories

Precisely, consider the partition $\{C_1, C_2, ..., C_n\}$ of the set C of distinctive features (see FP)[5]: C_1 contains such categories as consonantal, vocalic, sonorant, etc.; C_2 contains such categories as coronal, front, low, etc.[6], etc. (see FP). To the sequence $\{C_1, C_2, ..., C_n\}$ we add the set $C_0 = \{unit, segment\}$, where 'unit' and 'segment' are two categories not contained in C. Then, a phonological structural analysis of FP can be described as follows. Each segment or juncture *is a* $\langle 1, \text{unit} \rangle$, written $\langle 1, U \rangle$.[7] A maximal word-internal sequence of 'nodes' $\langle 1, U \rangle$ *is a* $\langle 2, U \rangle$. Consider now 'segment', written S. Each segment *is a* $\langle 1, S, U \rangle$. Let us define the notion 'x-continuous', where x is a complex category: a sequence of nodes a, where a is a complex category, is x-continuous in the representation of a word w iff. w cannot be analyzed as $\cdots \varphi \cdots X \cdots \psi \cdots$, where φ *is a* a, ψ *is a* a, X *is a* x, and φ and ψ correspond to two consecutive nodes in the sequence. Then a maximal $\langle 1, U \rangle$-continuous sequence of nodes $\langle 1, S, U \rangle$ *is a* $\langle 2, S, U \rangle$. A maximal word-internal sequence of nodes $\langle 2, S, U \rangle$ *is a* $\langle 3, S, U \rangle$. The node $\langle 3, S, U \rangle$ is directly dominated by $\langle 2, U \rangle$: we say that S is 'subordinate to' U.

Consider now C_1. Let us define the set $D(C_1) = D_1$: x belongs to $D(C_1) = D_1$ iff. $x = \alpha_1 c_1 \wedge \alpha_2 c_2 \wedge \cdots \wedge \alpha_p c_p$, $p \geqslant 1$, $c_i \in C_1$ and $\alpha_i \in \{+, -\}$ for $1 \leqslant i \leqslant p$, $c_h \neq c_k$ for $1 \leqslant h < k \leqslant p$ (for '$\wedge$', read 'and'). Let d be a category in D_1. A string that belongs to the category d *is a* $\langle 1, d, S, U \rangle$. A maximal $\langle 1, U \rangle$-continuous sequence of nodes $\langle 1, d, S, U \rangle$ *is a* $\langle 2, d, S, U \rangle$. A maximal word-internal sequence of nodes $\langle 2, d, S, U \rangle$ *is a* $\langle 3, d, S, U \rangle$. Each category a in D_1 is subordinate to S: for a in D_1, $\langle 3, a, S, U \rangle$ is directly dominated by $\langle 3, S, U \rangle$.

Consider next C_2. $D_2 = D(C_2)$ is defined as above. For b in D_2, let us define the set $D_1(b)$: x belongs to $D_1(b)$ iff. x belongs to D_1 and b is subordinate to x. By construction, $D_1(b) \neq \varnothing$ for every b in D_2. Note that $D_1(b)$ may be strictly included in D_1 (see FP). The following properties hold (see FP):

(i) $D_1(+c) = D_1(-c)$, for every c in C_2

(ii) let $b = \alpha_1 c_1 \wedge \alpha_2 c_2 \wedge \cdots \wedge \alpha_q c_q$ be a category in D_2; then $D_1(b) = D_1(\alpha_1 c_1) \cap D_1(\alpha_2 c_2) \cap \cdots \cap D_1(\alpha_q c_q)$

(iii) if $D_1(b)$ contains a, then $D_1(b)$ contains $a \wedge x$, where $a \wedge x$ belongs to D_1

Let e be a category in D_2 and d, a category in $D_1(e)$. A string that belongs to the category $e \wedge d$ *is a* $\langle 1, e, d, S, U \rangle$. A maximal $\langle 1, d, S, U \rangle$-continuous sequence of nodes $\langle 1, e, d, S, U \rangle$ *is a* $\langle 2, e, d, S, U \rangle$. A maximal word-internal sequence of nodes $\langle 2, e, d, S, U \rangle$ *is a* $\langle 3, e, d, S, U \rangle$. $\langle 3, e, d, S, U \rangle$ is directly dominated by $\langle 3, d, S, U \rangle$.

In a similar fashion, we define complex categories of the form $\langle i, f, e, d, S, U \rangle$, $i = 1$ or 2 or 3, $f \in D_3 = D(C_3)$, $e \in D_2(f)$, $d \in D_1(e)$, and, more generally, of the form $\langle i, h_m, h_{m-1}, ..., h_1, S, U \rangle$, $i = 1$ or 2 or 3, $m \leqslant n$, $h_m \in D_m = D(C_m)$, $h_l \in D_l(h_{l+1})$, $1 \leqslant l \leqslant m - 1$ (see FP).

A phonological structural analysis, as defined above, may be represented as a set of trees in which the leaves are terminal symbols (segments or junctures), and the other nodes are labelled with complex categories $\langle k, ..., U \rangle$. Alternatively, the same information can be represented by a family of labelled bracketings.

To illustrate the phonological representation of the form 'ppapapaapppapa' includes the structural analysis in

(1)   (where 'p' and 'a' have their usual phonetic meaning, and where 'C^i' stands for $\langle i, -\text{syllabic}, S, U \rangle$, '$V^i$', for $\langle i, +\text{syllabic}, S, U \rangle$, and '$R$', for $\langle 3, S, U \rangle$):

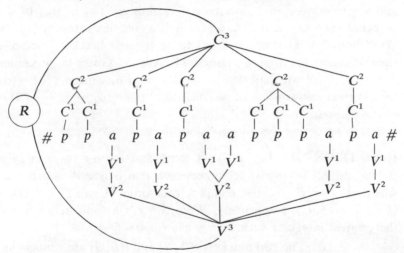

We reformulate the definition of the notion 'word' (see Footnote 4) as in (2):

(2)   A word is a terminal string which *is a* $\langle 2, U \rangle$

Each word w is provided with a phonological structural analysis $P(w)$. Call the pair $(w, P(w))$ a P-word. Characteristically, certain discontinuous sequences of substrings of w are analyzed as complex categories in the P-word $(w, P(w))$ (e.g. the sequence of vowels in (1), which *is a*

V^3, or the sequence of consonants, which *is a* C^3). More specifically, two consecutive nodes in a tree of $P(w)$ may dominate two non-adjacent substrings of w. Below we give a formal account of this characteristic property.

In what follows we will consider a word to be a string of indexed terminal symbols: the indices on the terminal symbols mark their positions within the word. Thus 'apap' will be written $\#_1 a_2 p_3 a_4 p_5 \#_6$. A *subword* of a word w is a string of indexed terminal symbols which is a substring of w. Thus $a_2 p_3$ is a subword of 'apap'.[8] Note that $a_2 p_3$ *and* $a_4 p_5$ are two different subwords of 'apap'. Thus, in such a formula as $w = \cdots \varphi \cdots$, where φ is a given subword of w, '$\cdots\cdots$' is always unambiguously interpreted. For a given word w, let us define a w-*sequence* to be an ordered m-tuple $(\psi_1, ..., \psi_m)$ of subwords of w such that, for each $i = 1, ..., m$, $\psi_i \neq \varnothing$ (where '$\varnothing$' is the empty word) and $w = \cdots \psi_i \varphi_i \psi_{i+1} \cdots \cdots$, where $\varphi_i \neq \varnothing$. Let $\alpha = (\psi_1, ..., \psi_m)$ be a w-sequence such that α *is a* A, where A is a complex category $\langle i, ..., U \rangle$. We define $l(A, \alpha)$ (resp. $r(A, \alpha)$) to be the left bracket $[_A$ (resp. the right bracket $]_A$) that is associated with α. Given a w-sequence $\alpha = (\psi_1, ..., \psi_m)$, we shall define *the closure of* α, written $c(\alpha)$: $c(\alpha)$ is the shortest subword v of w such that $v = \cdots \psi_1 \cdots \psi_m \cdots$. Clearly, $c(\alpha) = \psi_1 \cdots \psi_m$.

Below we define the notion 'contiguous'.

DEFINITION 2.1.   Let w be a word and let $\alpha = (\psi_1, ..., \psi_m)$ and $\beta = (\varphi_1, ..., \varphi_n)$ be two w-sequences such that α *is a* A and β *is a* B, where A and B are the complex categories $\langle i, g_p, ..., g_1, S, U \rangle$ and $\langle j, h_q, ..., h_1, S, U \rangle$, respectively. Write $S = g_0 = h_0$ and define s to be the greatest integer r such that $g_f = h_f$ for $0 \leq f \leq r$.

   (i)   The two brackets $r(A, \alpha)$ and $l(B, \beta)$ are *contiguous* iff. $w = \cdots \psi_m \cdots \varphi_1 \cdots$ and w is   different   from $\cdots \psi_m \cdots \chi \cdots \varphi_1 \cdots$, where χ is a $\langle 1, g_s, ..., g_0, U \rangle$ (equivalently a $\langle 1, h_s, ..., h_0, U \rangle$)

   (ii)   The two left brackets $l(A, \alpha)$ and $l(B, \beta)$ are *contiguous* iff. either (a) there exists an index t such that $\varphi_t = \cdots c(\alpha) \cdots$ and $c(\beta)$ is different from $\cdots \chi \cdots \psi_1 \cdots$, where χ is a $\langle 1, g_p \cdots, g_0, U \rangle$, or (b) there exists an index t such that $\psi_t = \cdots c(\beta) \cdots$ and $c(\alpha)$ is different from $\cdots \chi \cdots \varphi_1 \cdots$, where χ is a $\langle 1, h_q, ..., h_0, U \rangle$. Symmetrically for the right brackets $r(A, \alpha)$ and $r(B, \beta)$.

To illustrate, $r(V^1, a_4)$ and $l(V^1, a_6)$ in (1) are contiguous, as well as $r(V^1, a_4)$ and $l(C^2, p_5)$ (where x_i is the ith unit in the word). Similarly, $l(V^1, a_4)$ and $l(\langle 2, U \rangle)$ are contiguous.

3. THE GENERAL FORM OF PHONOLOGICAL RULES

3.1. In this paper we assume the FP theory of phonological representations. We will be interested in the part of the phonological component which maps the level of systematic phonemics onto the level of systematic phonetics. This mapping is effected by an ordered set of transformational rules which apply in sequence. A subset of rules apply cyclically to successively more dominant constituents of the surface structure. The other rules are word-level rules: they only apply to constituents which are words.

We will limit our discussion to word-level rules. Each such rule is defined in terms of a *structural description* (SD) and a *structural change* (SC). The SD specifies the domain of the transformation and the SC states what effect the transformation has on an arbitrary member of this domain. The domain of a transformation is a set of P-words (i.e. a set of families of labelled bracketings). A rule applies to a P-word $(w, P(w))$ which is analyzed into a set of 'projections':

DEFINITION 3.1. A labelled bracketing φ is *a projection of* $(w, P(w))$ if

(i) $\varphi = l(A, \alpha) a r(A, \alpha)$, where a is a non-empty subword of w which *is a* A, and α is the w-sequence (a), or

(ii) $\varphi = \psi\omega$, where ψ and ω are projections of $(w, P(w))$ such that $\psi = \cdots r(B, \beta)$, $\omega = l(C, \gamma) \cdots$, and $r(B, \beta)$ and $l(C, \gamma)$ are contiguous, or

(iii) $\varphi = l(D, \delta)\psi r(D, \delta)$, where ψ is a projection of $(w, P(w))$ such that $\psi = l(M, \mu) \cdots r(N, \nu)$, $\mu \neq \nu$, and $l(M, \mu)$ (resp. $r(N, \nu)$) and $l(D, \delta)$ (resp. $r(D, \delta)$) are contiguous.

An *exhaustive projection of* $(w, P(w))$ is a projection of $(w, P(w))\varphi$ such that $\varphi = l(\langle 2, U \rangle, w) \cdots r(\langle 2, U \rangle, w)$. Note that a projection is a well-formed labelled bracketing (its brackets occur in nested matched pairs). Each projection φ admits one or more (standard) factorizations, which are n-tuples of labelled bracketings $(\varphi_1, ..., \varphi_n)$ that satisfy the conditions (i) $\varphi = \varphi_1 \cdots \varphi_n$ and (ii) for each $i = 1, ..., n$, the leftmost symbol of φ_i is not a right bracket, nor is the rightmost symbol a left bracket.

To illustrate, $(\varphi_1, \varphi_2, \varphi_3, \varphi_4, \varphi_5)$ in (3) is a factorization of an exhaustive projection of the P-word in (1):

(3) $(\varphi_1, \varphi_2, \varphi_3, \varphi_4, \varphi_5)$ where

$\varphi_1 = l(\langle 2, U \rangle) l(V^1, a_4) a_4 r(V^1, a_4)$

$\varphi_2 = l(V^2, a_6) a_6 r(V^2, a_6)$

$\varphi_3 = l(C^1, p_7) p_7 r(C^1, p_7)$

$\varphi_4 = l(C^2, p_{10}p_{11}p_{12}) p_{10}p_{11}p_{12} r(C^2, p_{10}p_{11}p_{12})$

$\varphi_5 = l(C^2, p_{14}) p_{14} r(C^2, p_{14}) r(\langle 2, U \rangle)$

Other factorizations are

(4) $(\varphi_1, \varphi_2\varphi_3, \varphi_4, \varphi_5)$, and

(5) $(\varphi_1, \varphi_2\varphi_3, \varphi_4\varphi_5)$, where φ_i are as in (3)

Because we allow e, the labelled bracketing of length 0, as a possible factor, the projection $\varphi_1\varphi_2\varphi_3\varphi_4\varphi_5$ above has infinitely many additional factorizations (e.g. $(e, \varphi_1\varphi_2\varphi_3\varphi_4\varphi_5)$).

The SD specifies the domain of a transformation by placing conditions on factorizations of exhaustive projections: a P-word x meets the SD iff. there exist an exhaustive projection of $x\,y$ and a factorization of $y\,z$ such that z meets the SD. Precisely, an n-term SD is an n-ary compound predicate constructed from a set of n-ary basic predicates and the connectives '$\wedge$' and '$\vee$'. For every complex category $C = \langle i, ..., U \rangle$ and integers $n, p, q, [C]^n_{p-q}$ is a basic predicate; $[C]^n_{p-q}$ is true of a factorization $(\psi_1, ..., \psi_n)$ if $1 \leq p \leq q \leq n$ and $\psi_p \cdots \psi_q = \varphi l(C, \alpha) \cdots r(C, \alpha)\omega$ for some α, where $\varphi\omega$ is a string of brackets or e. Let us define the *debracketing function* d: $d(x)$ is x if x is a terminal symbol and is e if x is a bracket; $d(\varphi\psi) = d(\varphi)d(\psi)$. For every specified distinctive feature F and integers n, p, q, $[F]_{p-q}^n$ is a basic predicate; $[F]_{p-q}^n$ is true of a factorization $(\psi_1, ..., \psi_n)$ if $1 \leqslant p \leqslant q \leqslant n$ and $d(\psi_p \cdots \psi_q)$ belongs to F. Let $\mathcal{F}$ be a set of specified distinctive features, $\mathcal{F} = \{F_j\}_{j \in J}$; we will write $\bigwedge_{j \in J} [F_j]_{p-q}^n = [\mathcal{F}]_{p-q}^n$. We shall also define the basic predicate Φ_p^n and the predicate E_p^n (E_p^n is not a basic predicate): Φ_p^n is true of a factorization $(\psi_1, ..., \psi_n)$ if $1 \leqslant p \leqslant n$ and $\psi_p = e$; E_n^p is true of a factorization $(\psi_1, ..., \psi_n)$ if $1 \leqslant p \leqslant n$ and $\psi_p \neq e$. For $\bigwedge_{p \leqslant j \leqslant q} E_j^n$ (resp. $\bigwedge_{p \leqslant j \leqslant q} \Phi_j^n$), where $p < q$, we will write E_{p-q}^n (resp. Φ_{p-q}^n). For every basic predicate P_{p-q}^n (resp. P_{p-p}^n) the following axioms hold (see FP): $P_{p-q}^n \equiv P_{p-q}^n \wedge E_{p-q}^n$ (resp. $P_{p-p}^n \equiv P_{p-p}^n \wedge E_p^n$), and $P_{p-q}^n \wedge \Phi_{p-q}^n \equiv \Phi_{p-q}^n$ (resp. $P_{p-q}^n \wedge \Phi_p^n \equiv \Phi_p^n$) (where '$\equiv$' means 'equivalent to'). Furthermore: $E_p^n \wedge \Phi_p^n \equiv \Phi_p^n$.

The following examples illustrate these notions.

Example 1. Sanskrit [n] – retroflexion

A coronal nasal when immediately followed by a sonorant is turned into a retroflex nasal if preceded in the same word by a retroflex coronal continuant non-syllabic, provided that no coronal non-syllabic intervenes (see Whitney, 1941). The SD is:

$$[_A vbl, [_B \alpha]_B, [_B \beta]_B, [_c \gamma]_c, vbl]_A$$
$$\quad 1 \quad\quad 2 \quad\quad 3 \quad\quad 4 \quad\quad 5$$

where A is $\langle 2, U \rangle$, B is $\langle 1, +\text{coronal}, -\text{syllabic}, S, U \rangle$, C is $\langle 1, U \rangle$. α is the feature matrix (+continuant, +retroflex), β is the feature matrix (+nasal), and γ is the feature matrix (+sonorant). For vbl, read 'variable'.

Example 2. Menomini vowel raising

A long non-low vowel is turned into a high vowel if it is followed in the same word by a postconsonantal high glide or a high vowel (see Bloomfield, 1939). The SD is:

$$[_A vbl, B, vbl, C]_A$$
$$\quad 1 \quad 2 \quad 3 \quad 4$$

where A is $\langle 2, U \rangle$, B is $\langle 1, +\text{long}, -\text{low}, \_\text{syll}, S, U \rangle$, C is $\langle 1, +\text{high}, +\text{syll}, S, U \rangle$. Note that a postconsonantal glide is (+syllabic, −vocalic).

Example 3.   *Old Norse umlaut*

A vowel or a diphthong is fronted if it is followed in the same word by [i] or [j], provided that only non-syllabics intervene (see Kiparsky, 1973). The SD is:

$$[_A \, vbl, B, [_C\alpha]_C, vbl \, _A]$$
$$\quad 1 \quad\ 2 \quad 3 \qquad 4$$

where A is $\langle 2, U \rangle$, B is $\langle 2, +\text{syll}, S, U \rangle$ and C is $\langle 1, (+\text{syll}), S, U \rangle$[9]. α is the feature matrix (+high, +front, −cons).

Example 4.   *Russian voicing assimilation*

In obstruent clusters the voiced/voiceless quality of the final obstruent governs the voicing of all obstruents in the cluster (see Lightner, 1972). The SD is:

$$[_A vbl, [_B vbl, [_C\alpha]_C {}_B], vbl]_A$$
$$\quad 1 \qquad 2 \qquad 3 \qquad\ \ 4$$

where A is $\langle 2, U \rangle$, B is $\langle 2, -\text{sonorant}, S, U \rangle$, C is $\langle 1, U \rangle$. α is the feature matrix (±voice).

3.2.   The SD may only include disjunctions of a very restricted type. Specifically, it may include *D-predicates*, defined in terms of the following predicates[10]:

(6)   (i) (E_p^n): $(E)_p^n$ is the predicate $(\Phi_p^n \vee E_p^n)$

        (ii) $([F]_{p-q}^n)$, where F is a specified feature: $([F]_{p-q}^n)$ is the predicate $([-F]_{p-q}^n \vee [F]_{p-q}^n)$

        (iii) $(E_p^n \supset E_q^n)$: $(E_p^n \supset E_q^n)$ is the predicate $(\Phi_p^n \vee E_q^n)$

        (iv) $(E_r^n \supset [F]_{p-q}^n)$, where F is a specified feature: $(E_r^n \supset [F]_{p-q}^n)$ is the predicate $(\Phi_r^n \vee [F]_{p-q}^n)$

        (v) $([F]_{p-q}^n \supset E_r^n)$, where F is a specified feature: $([F]_{p-q}^n \supset E_r^n)$ is the predicate $([-F]_{p-q}^n \vee E_r^n)$

        (vi) $([F]_{p-q}^n \supset [G]_{r-s}^n)$, where F and G are specified features: $([F]_{p-q}^n \supset [G]_{r-s}^n)$ is the predicate $([-F]_{p-q}^n \vee [G]_{r-s}^n)$

For $(E_{j_0}^n) \wedge \left(\bigwedge\limits_{j,k \in J} (E_j \supset E_k) \right)$, where $j_0 \in J$, we will write $\left(\bigwedge\limits_{j \in J} E_j^n \right)$. A *D-predicate* is defined as follows:

DEFINITION 3.2.   A disjunction of conjunctions of basic predicates is a *D-predicate* if it is equivalent to a predicate D such that

> (i) D is a conjunction of predicates of types (6i)–(6vi)
>
> (ii) if $D \equiv (x \supset y) \wedge \cdots$ or $D \equiv (y \supset x) \wedge \cdots$, then $D \equiv (x) \wedge \cdots$

A SD can be defined to be a conjunction of basic predicates and/or D-predicates.

3.3.   Suppose that we impose the general condition (7): (7) For any transformation T_i and any P-word P_j in the domain of T_i, P_j must be unambiguously analyzed by the SD of T_i. See FP for justifications. Let us assume the notion *general variable*: the ith factor of a SD is a general variable if it is a *vbl* (i.e. an arbitrary factor) and if $[\langle 2, U \rangle]_{n-n}^n$ is the only predicate of the form $[C]_{p_i - q_i}^n$, where C is a complex category and $p_i \leqslant i \leqslant q_i$. A well-formed SD must include at least one general variable; see FP. From condition (7) it follows that a SD may include at most one general variable. Thus, the rules in the examples of 3.1, for instance, must be formulated in a different way. To accommodate such cases will introduce expressions of the form $\mathscr{C}_x = \mathscr{C}' \wedge X_{i-j}^n$, $i = 1$ or $j = n$, where $\mathscr{C}'$ is an n-term SD which incorporates no predicate P_{k-l}^n (resp. P_k^n) with $i \leqslant k \leqslant j$ and/or $i \leqslant l \leqslant j$ (resp. $i \leqslant k \leqslant j$), and X_{i-j}^n is an auxiliary symbol: $\mathscr{C}_x$ is interpreted as an abbreviation for the set of structural descriptions $\{\mathscr{C}_p\}_{p \geqslant 0}$, where $\mathscr{C}_p$ is the predicate $\mathscr{C}' \wedge [\langle 1, U \rangle^p]_{i-j}^n$ (the predicate $[\langle 1, U \rangle^p]_{i-j}^n$ is true of a factorization $(\psi_1, ..., \psi_n)$ if $1 \leqslant i \leqslant j \leqslant n$ and $\psi_i \cdots \psi_j$ is analyzed as a sequence of p nodes $\langle 1, U \rangle$). Correspondingly, the expression $T_x = (\mathscr{C}_x, M)$, where M is a SC, is interpreted as an abbreviation for the set of transformations $\{T_p = (\mathscr{C}_p, M)\}_{p \geqslant 0}$. We will call the auxiliary symbol 'X_{i-j}^n' a *parametric variable* and T_x, a *schema*. More generally, we will define a schema to be an expression which is constructed from the symbols that appear in transformations and certain auxiliary expressions defined by the theory. The theory specifies certain rules of expansion that eliminate auxiliary symbols. A *well-formed schema* is defined to be a schema which can be expanded, in the manner provided, into a set of transformation. Assuming these conventions we can reformulate the rules in the examples of 3.1 as schemata, with X_{i-j}^n replacing one of the two general variables. For example, we can write the SD of the Russian voicing assimilation rule as in (8):

(8) $[_A X, [_B vbl, [_C \alpha]_C]_B, vbl]_A$
       1   2   3   4

where A, B, C, and α are as in example 4 of 3.1, and where '$(X, \psi_2, \psi_3, \psi_4)$' is the auxiliary 'predicate' X_{1-1}^4

It is clear that the transformations in the expansion of (8) conform to condition (7).

To illustrate further the above principles we formulate in (9) the SD of the rule of English which lengthens non-low vowels in prevocalic or final position (see SPE):

(9) $[_A X, [_B \alpha]_B, [_B \beta]_B, vbl]_A$
       1   2   (3)   (4)

Condition: $(4 \supset 3)$, where A is $\langle 2, U \rangle$, B is $\langle 1, U \rangle$, α is the feature matrix $(+\text{syll}, -\text{low})$, and β is the feature matrix $(+\text{syll})$; '(i)', $i = 3$ or 4, is the predicate (E_i^4), and '$(4 \supset 3)$' is the predicate $(E_4^4 \supset E_3^4)$

Let $\mathscr{C}_x$ be the SD in (9). $\mathscr{C}_x$ abbreviates the set $\{\mathscr{C}_p\}_{p \geqslant 0}$ where $\mathscr{C}_p$ is derived from (9) by replacing 'X' by '$\langle 1, U \rangle^p$'! Consider the form *rodeo*. It meets $\mathscr{C}_3$ and $\mathscr{C}_4$. Consequently, the last two vowels of *rodeo* are lengthened. The application of a schema $T_x = (\mathscr{C}_x, M)$ to a form is governed by the following algorithm:

(10)   Let $\{T_h\}_{h \in H}$ be the expansion of T_x and let $\mathscr{P}$ be a P-word. For each $h \in H$, if $\mathscr{P}$ meets the SD of T_h mark with an ε the subword of $\mathscr{P}$ which is to be changed by T_h. Apply the SC simultaneously to all epsilonized elements.

Note that no schema may contain more than one parametric variable (see FP).

4. THE PARENTHESIS NOTATION AND ALTERNATING PATTERNS

4.1. *The Parenthesis Notation*

4.1.1. Linguistic theory includes a metric which assigns a numerical measure of valuation to each grammar. In the standard theory the value of a grammar is its length, in terms of number of symbols. This measure is based on a system of abbreviatory devices which, in certain limited and well-defined circumstances, effect the collapsing of formally related rules. An important element of this system of conventions in the SPE theory is the parenthesis notation. Briefly, this notation states that two consecutive rules of the form PQR and PR collapse into a single schema of the form $P(Q)R$; similarly two consecutive rules of the form $P[\alpha F]Q$ and $P[-\alpha F]Q$, where F is a distinctive feature and α is + or −, collapse into a single schema of the form $P([\alpha F])Q$. Rules so collapsed are disjunctively ordered (see below). The parenthesis notation has different realizations in the SPE theory and in the FP theory, due essentially to differences in the ways in which these theories define the notion 'schema'. Whereas the parenthesis notation is exclusively a collapsing device in the SPE theory (necessarily associated with disjunctive ordering), it has other functions as well in the FP theory: for example, it provides the formal representation for $\cup$-predicates (see Section 3.2; in FP, D-predicates are elements of transformations; they are elements of schemata only derivatively). Note however that the theoretical notions associated with the parenthesis notation have the same extension in both the SPE theory and the FP theory. In the present section we will consider the abbreviatory function of the parenthesis notation within FP and we will examine its relation to disjunctive ordering. The distinctive property of 'disjunctive' schemata is stated in (11):

(11)   Let $S = \{T_1, T_2, ..., T_m\}$ be a set of transformations that are collapsed into a schema by the parenthesis notation. Then, any two elements T_i, T_j of S have different structural changes.

The parenthesis notation is defined over a strict subset of transformations. For the purpose of our discussion we may assume that the

class of relevant transformations is partitioned into three subclasses:
the class of deletion transformations, the class of insertion transformations, and the class of copying transformations. The SC of an n-term
deletion transformation T is an expression $M_d^n(i)$, where $1 \leqslant i \leqslant n$; if
the ith factor is a vbl, $T = (\mathscr{C}, M_d^n(i))$ (where $\mathscr{C}$ is the SD of T) is
undefined; otherwise, the effect of T is to delete the ith factor. The SC
of an n-term insertion transformation T' is an expression $M_e^n(i, \mathscr{F})$,
where $\mathscr{F}$ is a set of specified distinctive features and $1 \leqslant i \leqslant n$; if the ith
factor is a vbl, $T' = (\mathscr{C}', M_e^n(i, \mathscr{F}))$ is undefined; otherwise, the effect of
T' is to substitute the set of specified features $\mathscr{F}$ for the corresponding
set in the ith factor, or for the ith factor if it is e. The SC of an n-term
copying transformation T'' is an expression $M_c^n(i, j)$ or $M_c^n(i, j, \mathscr{N}, \mathscr{P})$,
where $\mathscr{N}$ is a set of p (unspecified) distinctive features $\{F_{h_k}\}_{1 \leqslant k \leqslant p}$ such
that $\mathscr{C}''$, the SD of T'', contains $[\alpha_k F_{h_k}]_{j-j}^n$, $\alpha_k = +$ or $-$ for $1 \leqslant k \leqslant p$,
where $\mathscr{P}$ is a p-tuple $(q_1, ..., q_p)$, $q_k = +$ or $-$ for $1 \leqslant k \leqslant p$, and where
$1 \leqslant i \leqslant n, 1 \leqslant j \leqslant n$; if the jth factor is e or vbl, or if the ith factor is vbl,
T'' is undefined; otherwise, the effect of $(\mathscr{C}'', M_c^n(i, j, \mathscr{N}, \mathscr{P}))$ is to substitute the set of specified features $\{q_k \alpha_k F_{h_k}\}_{1 \leqslant k \leqslant p}$ (where α_k is as defined
above) for the corresponding set in the ith factor (or for the ith factor
if it is e), and the effect of $(\mathscr{C}'', M_c^n(i, j))$ is to substitute a copy of the jth
factor for the ith factor.

The transformational mappings defined above exclusively affect the
phonological matrix (the terminal string of the P-word). The required
modifications in the labelled bracketings are performed by universal
(automatic) conventions; see FP.

We shall define the function v which maps structural changes into
abstract predicates.

DEFINITION 4.1.   Let $T = (\mathscr{C}, M)$ be an n-term transformation,
where $\mathscr{C}$ is the SD and M, the SC. The *value of M*, $v(M)$, is an abstract
n-ary predicate. v is defined only if T is defined. If $M = M_d^n(i)$, then
$v(M)$ is the abstract predicate Φ_0^n. If $M = M_e^n(i, \mathscr{F})$, $v(M)$ is $[\mathscr{F}]_0^n$. If
$M = M_c^n(i, j, \mathscr{N}, \mathscr{P})$, where $\mathscr{N} = \{F_{h_k}\}_{1 \leqslant k \leqslant p}$, $\mathscr{C}$ contains $[\alpha_k F_{h_k}]_{j-j}^n$, $\alpha_k = +$
or $-$, for $1 \leqslant k \leqslant p$, $\mathscr{P}$ is $(q_1, ..., q_p)$, $q_i = +$ or $-$, then $v(M) =$
$[\{q_k \alpha_k F_{h_k}\}_{1 \leqslant k \leqslant p}]_0^n$. If $M = M_c^n(i, j)$, $v(M)$ is E_0^n. In all other cases $v(M)$ is
undefined.

The following relations hold: $P_0^n \equiv P_0^n \wedge E_0^n$, and $\Phi_0^n \wedge P_0^n \equiv \Phi_0^n$, for every predicate P_0^n (see FP). We extend the mapping v in the following way:

(12) Let $T = (\mathscr{C}, M)$ be a transformation. The value of T, $v(T)$, is defined to be the predicate $\mathscr{C} \wedge v(M)$.

Let $\mathscr{T} = \{T_f\}_{1 \leqslant f \leqslant m}$ be a set of n-term transformations such that for any two members T_g and T_h of $\mathscr{T}$, if the SC of T_g is $M_d^n(i)$, or $M_e^n(i, \mathscr{F})$, or $M_c^n(i, j, \mathscr{N}, \mathscr{P})$, or $M_c^n(i, j)$, and the SC of T_h is $M_d^n(k)$, or $M_c^n(k, l, \mathscr{N}', \mathscr{P}')$, or $M_c^n(k, l)$, then $i = k$ and $j = 1$. Let us write $T_f = (\mathscr{C}_f, M_f)$ for T_f in $\mathscr{T}$. Then, if the predicate $\bigvee_{1 \leqslant f \leqslant m} v(T_f)$ is a D-predicate, $\mathscr{T}$ coincides with the expansion of the schema $\Sigma = (\Gamma, \Delta, \Lambda)$, where $\Gamma = \bigvee_{1 \leqslant f \leqslant m} \mathscr{C}_f$, $v(\Delta) = \bigvee_{1 \leqslant f \leqslant m} v(M_f)$, and $v(\Sigma) = \bigvee_{1 \leqslant f \leqslant m} v(T_f) = \Gamma \wedge v(\Delta) \wedge \Lambda$. The rules in the expansion of Σ are *disjunctively ordered*:

(13) (i) Let T_p and T_q be two transformations in the expansion of Σ. If $v(T_q) \equiv \Phi_i^n \wedge \cdots$ and $v(T_p) \not\equiv \Phi_i^n \wedge \cdots$, then T_p is ordered before T_q

(ii) Let $\mathscr{P}$ be a P-word in the domain of Σ and let s be a segment of $\mathscr{P}$. At most one transformation in the expansion of Σ may affect s.

We revise the algorithm (10) to (14):

(14) Let Σ_x be a schema which includes a parametric variable X. The elimination of X precedes the elimination of the parenthesis notation. The expansion of Σ_x by the parametric variable yields a set of schemata $\{\Sigma_p\}_{p \geqslant 0}$. For every p the expansion of Σ_p by the parenthesis notation yields a set of transformations $\{T_{i,p}\}_{1 \leqslant i \leqslant m}$ (where possibly $m = 1$). Let $\mathscr{P}$ be a P-word. For each $p \geqslant 0$, if $\mathscr{P}$ meets the SD of Σ_p, mark with ε the segments of $\mathscr{P}$ which are to be changed by Σ_p. Apply the structural changes simultaneously to all epsilonized elements.

To illustrate, the schema in (15) represents a process by which a word final consonant is assimilated to an initial consonant in the following word and deleted elsewhere:

(15)   SD: $X, \alpha, [_A \beta]_A, \alpha, vbl$
              1 2   3 $(4)_a$· 5

         SC: 1 $(4)_b$ 3 $(4)_a$ 5
         Condition: $a \equiv b$
         where A is $\langle 1, U \rangle$, α is $\langle 1, -\text{syll}, S, U \rangle$, β is #, '$(4)_a$'
         represents the predicate (E_4^5), '$(4)_b$', the predicate (E_0^5), and
         '$a \equiv b$', the predicate $(E_0^5 \equiv E_4^5)$.

4.1.2.   Consider the Latin stress rule: stress falls on the antepenultimate syllable if one exists and if the penultimate syllable is weak; stress falls on the only vowel in monosyllables and is penultimate under all other conditions. The SD of this rule is formulated in (16) (irrelevant details omitted):

(16) $vbl, V, \check{V}, C, V, \#$
         1   2 (3)(4)(5) 6

         Conditions: $(4 \supset 3)$, $(3 \supset 5)$
         where V is $\langle 1, +\text{syll}, S, U \rangle$, $\check{V}$, $\langle 1, -\text{long}, +\text{syll}, S, U \rangle$, and
         C, $\langle 1, -\text{syll}, S, U \rangle$

The effect of the SC associated with (16) is to build up one of the following metric trees (where 'S' means 'strong' and 'W', 'weak', see M. Liberman, 1975):

(17)   (i)   (ii)   (iii)

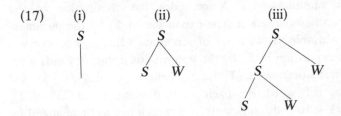

(17i) corresponds to the SD '1, 2, 6', (17ii), to the SD '1, 2, 5, 6', and

(17iii), to the SD '1, 2, 3, (4), 5, 6'. The SC associated with (16) can actually be formulated as an operation which inserts the brackets ']ₛ[_w' to the right of the second factor in (16) (see FP). The combination of this SC and of (16) is a 'disjunctive' schema.

4.2. *Alternating Patterns*

Consider the following SD:

(18)   # , X, V, V, vbl
        1 2 (3) 4 5

    where V is ⟨1, +syll, S, U⟩

The schema in (18) expands into the following family of SD's:

(19)   #, ⟨1, U⟩ᵖ, V, V, vbl
        1   2   (3) 4 5

    p=0, 1, 2, ···

Suppose that to each SD in (19) a SC is associated which builds up one of the metric trees in (20) (along the lines of 4.1.2):

(20)   (i)   (ii)

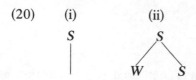

((20i) is built up only if the input cannot be analyzed as '1, 2, 3, 4, 5'.) The transformation associated with (18) will then generate a sequence of such trees. The following well-formedness conditions must be met (see FP):

(21)   (i)   Each syllable must be dominated by one of the nodes *S* or *W*

        (ii) A syllable may not be dominated by more than one node.

From (21), it follows at once that only a subset of (19) can apply, namely the subset:

(22) $\#, \langle 1, U \rangle^{2p}, V, V, vbl$
      1   2   (3) 4   5

      $p = 0, 1, 2, \cdots$

Suppose S in (20) is associated with [+stress]. Consider the form in (23):

(23) $\# \; V \; V \; V \; V \; V \; V \; V \; V \; V \; \#$

when (18) applies to (23), the following output obtains (where '$\acute{V}$' is $\langle 1, +\text{stress}, +\text{syll}, S, U \rangle$):

(24) $\# \; V \; \acute{V} \; V \; \acute{V} \; V \; \acute{V} \; V \; \acute{V} \; \acute{V} \; \#$

when (18) applies to (25), (26) obtains:

(25) $\# \; V \; V \; V \; V \; V \; V \; V \; V \; \#$
(26) $\# \; V \; \acute{V} \; V \; \acute{V} \; V \; \acute{V} \; V \; \acute{V} \; \#$

We see that (18) generates the stress pattern of Northern Paiute. The stress pattern of Southern Paiute is generated by (27):

(27)   SD: $\#, X, V, V, V, vbl$
               1 2 (3) 4 5 6
         SC: 1 2 (3)[+ 5 6
                   stress]

Note that the lengthening rule in Tübatülabal can be written as follows ('strong' is associated with [−long], and 'weak', with [+long]):

(28)

         SD: $\#, X, V, \overset{(U)_b}{V}, vbl$ $(a \supset b)$
              1 2 (3)$_a$ 4 5
                   $\uparrow$
         SC: S
         where $\overset{(U)}{V}^b$ is $\langle 1, (-\text{long})_b, +\text{syll}, S, U \rangle$

Université de Paris VII

NOTES

* I gratefully acknowledge the help of the following people: Jay Keyser, Morris Halle, Alan Prince, and Elisabeth Selkirk. This work was supported by C.N.R.S. grant E.R.A. No. 247.

[1] Chomsky and Halle (1968).

[2] The set of conventions that Chomsky and Halle propose actually has the form of a theory of markedness. To the best of my knowledge, theirs is the first attempt at building a systematic account of 'naturalness' in phonology. For further developments of the theory of markedness, see M. L. Kean (1975).

[3] For example, linguistic theory contains the following rule: the boundary $\#$ is automatically inserted at the beginning and end of every string dominated by a major category, i.e. by one of the lexical categories N, A, V, Adv, or by a category such as S, NP, VP, which dominates a lexical category. There are language-specific rules which delete $\#$ in various positions. For a discussion of such rules and related issues, see E. O. Selkirk (1972).

[4] A word is an element of the form $\#\cdots\#$, where $\cdots$ contains no occurrence of $\#\#$.

[5] This partition is an element of universal grammar.

[6] See *SPE* for a definition of these features.

[7] See J. Bresnan (1975) for the notation. A complex category is an ordered pair $\langle i, M \rangle$, where i is the type of the category and M is the feature-matrix of the category.

[8] Note that a_3p_4 is not a subword of 'apap', nor is a_2p_5.

[9] $\langle 1, (+\text{syll}), S, U \rangle$ is the disjunction '$\langle 1, +\text{syll}, S, U \rangle$ or $\langle 1, -\text{syll}, S, U \rangle$'; see FP.

[10] The list in (6) is not exhaustive; see FP and footnote 9. However, it will suffice for our discussion.

BIBLIOGRAPHY

Bloomfield, L.: 1939, 'Menomini morphophonemics', in *Etudes phonologiques dédiées à la mémoire de M. le Prince Troubetzkoy*, Univ. of Alabama Press, 1964.

Bresnan, J.: 1973, 'Sentence Stress and Syntactic Transformations', in K. J. J. J. Hintikka *et al.* (eds.), *Approaches to Natural Language* (Synthese Library), D. Reidel Publishing Company, Dordrecht and Boston, 1973.

Bresnan, J.: 1975, 'On the Form and Functioning of Transformations', mimeographed.

Chomsky, N.: 1967, 'Some General Properties of Phonological Rules', *Language* **43**, No. 1.

Chomsky, N. and Halle, M.: 1968, *The Sound Pattern of English*, Harper and Row, New York.

Halle, M.: 1975, 'Confessio grammatici', *Language* **51**, No. 3.

Halle, M., Prince, A. S., and Vergnaud, J.-R.: 1975, *Formal Phonology*, M.I.T. ms., Cambridge.

Kean, M. L.: 1975, *The Theory of Markedness in Generative Grammar*, M.I.T. Ph.D. dissertation.

Kiparsky, P.: 1973, ' "Elsewhere" in Phonology', in S. R. Anderson and P. Kiparsky (eds.), *A Festschrift for Morris Halle*, Holt, Rinehart and Winston, Inc., New York, 1973.

Liberman, M.: 1975, *The Intonational System of English*, M.I.T. Ph.D. dissertation.

Lightner, T.: 1972, *Problems in the Theory of Phonology*, Vol. 1, Linguistic Research, Inc., Edmonton.

Prince, A. S.: 1974, 'McCawley on Formalization', in *Recherches linguistiques*, 3.

Prince, A. S.: 1975, *The Phonology and Morphology of Tiberian Hebrew*, M.I.T. Ph.D. dissertation.

Selkirk, E. O.: 1972, *The Phrase Phonology of English and French*, M.I.T. Ph.D. dissertation.

Selkirk, E. O.: forthcoming, *Phonology and Syntax: The Relation between Sound and Structure*.

Vergnaud, J.-R.: 1973, 'A review of Chapter 8 of SPE', M.I.T. ms., Cambridge.

Vergnaud, J.-R.: 1974, *Problèmes formels en phonologie générative*, Thèse de 3e cycle, Paris VII.

Whitney, W. D.: 1889, *Sanskrit Grammar*, Harvard Univ. Press, Cambridge.

INDEX OF NAMES*

Adams, E. W., 46
Anderson, A. R., 150

Bacon, F., 121
Bar-Hillel, M., 176, 177, 178, 186
Bar-Hillel, Y., 240
Belnap, N., 150
Berkeley, G., 220
Bernoulli, J., 118
Bernstein, S. N., 105
Bresnan, J., 269, 271, 278, 280, 317
Beth, E. W., 63
Bloomfield, L., 307
Blum, M., 205
Borgida, E., 176, 180, 181, 182, 183
Bos, U., 208
Boyle, R., 121
Braithwaite, R. B., 149
Brody, B., 163
Bromberger, S., 158
Bronowsky, Y., 241

Campbell, D. T., 5, 6
Carnap, R., 105, 113, 162, 163, 225, 240
Chaitin, G. J., 193, 209, 211, 216
Chang, C. C., 66, 70, 73, 75
Chellas, B., 246
Chomsky, N., 228, 240, 262, 263, 264, 266, 267, 269, 271, 299, 317
Church, A., 164, 200
Coffa, J. A., 157
Cohen, L. J., 145, 240
Cooper, R., 266, 281
Cresswell, M. J., 246

Daley, R., 207
Dalla Chiara Scabia, M. L., 34
Darley, J. M., 180
Davidson, D., 228, 229, 240, 246, 285
Dewey, J., 3
Domotor, Z., 105
Duhem, P., 122

Edwards, W., 167
Engels, F., 122

Fagot, R. F., 46
Feyerabend, P. K., 20, 21, 22
Fine, T. L., 105, 107, 108, 110, 111, 114, 115
Finetti, B. de, 105, 107, 113, 187, 218
Fischhoff, B., 167
Fishburn, P. C., 105
Fodor, J. A., 225, 240, 281
Freid, C., 9
Fuchs, P., 200

Gabbay, D., 240, 281
Gacs, P., 206, 211
Geist, V., 6
Giles, R., 35
Gill, J. T., 105, 108
Gilles, D., 114
Goodman, N., 217
Gorsky, D. P., 71
Greeno, J. G., 152, 164
Grice, H. P., 231–235, 240, 286, 287, 288
Groen, G. J., 45, 46, 47

Hacking, I., 114
Halle, M., 299, 301, 317
Hammerton, M., 176
Harper, W., 11
Harris, J. H., 123, 143
Hartley, D., 121
Hattiangatti, J. H., 143
Hempel, C., 37, 143, 149, 151, 152, 154, 156, 157, 160, 162, 163, 164, 165
Heyting, A., 198
Hilpinen, R., 123, 137, 142, 145
Hintikka, J., 48, 49, 63, 70, 137, 140, 142, 144, 145, 235, 236, 239, 240, 281
Hooke, R., 121

*Names listed in the bibliographies are not included in this index.

Hooker, C. A., 3, 7, 9, 10, 11, 13, 19, 20, 21, 22
Hurwicz, L., 91, 101

Jackendoff, R., 281, 283, 284
Jeffrey, R. C., 19, 153
Johnson-Laird, P., 281, 285

Kahneman, D., 167, 173, 174, 186, 187
Kant, I., 17
Kaplan, M. A., 105, 110, 112
Karttunen, L., 293, 297
Kasher, A., 240, 247
Katz, J. J., 225, 226, 240
Kean, M. L., 317
Keisler, H. J., 70, 73
Keynes, J. M., 105, 113
Kiparsky, P., 263, 309
Kolmogorov, A. N., 193, 194, 195, 216, 217
Koopman, B. O., 105, 113
Koopmans, T., 53
Kraft, C., 105, 108
Krantz, D., 107
Kueker, D. W., 66, 75
Kuhn, T., 22, 38, 143
Kyburg, H. E., 152

Lakatos, I., 10, 21, 145
Lakoff, G., 264, 266
Lakoff, R., 240
Langendoen, T. D., 226, 241
Latané, B., 180
Laudan, L., 143
Leibniz, G., 29, 30
Lenin, V. I., 122
LeSage, G., 121
Levi, I., 19
Levin, L. A., 205, 209, 210
Lewis, D., 233, 240, 246, 247, 248
Liberman, M., 314
Lichenstein, S., 167
Lorentz, K., 6
Luce, R. D., 105
Lyon, D., 176
Lyons, J., 267

Mach, E., 25
Makkai, M., 66, 75
Martin-Löf, P., 195, 197, 198, 204, 213, 214, 216
McCawley, J., 264, 265
Meinhardt, G., 208

Miller, D., 11, 123, 127, 128, 136, 137, 143, 145
Miller, G., 281, 285
Mises, R. von, 156, 200, 213–222
Montague, R., 91, 97, 226, 240, 265, 281, 292
Myhill, J., 281

Nagel, E., 149
Narens, L., 105
Niiniluoto, I., 140, 143, 144, 145, 164
Nisbett, R. E., 176, 180, 181, 182, 183
Nowak, L., 146

Oppenheim, P., 151, 160

Padoa, A., 63
Parsons, T., 266
Partee, B., 281
Peirce, C. S., 122, 143
Peters, S., 293, 296, 297
Pope, A., 239
Popper, K., 3, 6, 9, 22, 122, 123, 124, 126, 141, 143, 145, 149
Postal, P., 262, 263, 265, 285
Priestley, J., 121
Przełęcki, M., 35, 49, 51, 63, 89
Putnam, H., 240

Quine, W. V. O., 143, 149

Rantala, V., 53, 63, 66, 70, 93, 99
Reichenbach, H., 122, 161, 165
Roberts, F. S., 107
Robinson, R. E., 46
Ross, J., 278, 279, 280
Rudner, R., 19

Sadovsky, V. N., 52, 69
Salmon, W., 114, 151, 163, 164, 165, 185
Savage, L. J., 105, 113, 187
Schnorr, C. P., 198, 199, 200, 201, 203, 204, 205, 206, 208, 213, 214, 219
Scott, D., 105
Scriven, M., 152
Searle, J. R., 229, 240, 247, 248
Selkirk, E., 281, 317
Simon, H. A., 45, 46, 47, 50, 52, 63, 93, 102
Slovic, P., 167, 176
Smirnov, V. A., 52, 69, 79

Smith, A. H., 163
Sneed, J., 32, 35, 46, 56, 59
Solomonoff, R. J., 216
Stalnaker, R. C., 129, 145, 240
Suppes, P., 51
Svenonius, L., 66, 68, 75, 76

Tarski, A., 63, 229, 286
Thomason, R. H., 226, 240, 241
Tichý, P., 123, 143, 144, 145
Toraldo di Francia, F., 34
Tuomela, R., 46, 48, 49, 52, 63, 144, 145
Turoff, M., 173

Tversky, A., 167, 173, 174, 176, 186, 187

Ville, J., 200
Villegas, C., 107

Watkins, J., 123
Whewell, W., 122
Whitney, W. D., 307
Wilson, R., 6
Winnie, J., 89
Wójcicki, R., 34, 35, 37, 91
Wunderlich, P., 245

Zvonkin, A., 209

THE UNIVERSITY OF WESTERN ONTARIO
SERIES IN PHILOSOPHY OF SCIENCE

A Series of Books on Philosophy of Science, Methodology, and Epistemology published in connection with the University of Western Ontario Philosophy of Science Programme

Managing Editor:

1. J. Leach, R. Butts, and G. Pearce (eds.), *Science, Decision and Value.* Proceedings of the Fifth University of Western Ontario Philosophy Colloquium, 1969. 1973, vii + 213 pp.
2. C. A. Hooker (ed.), *Contemporary Research in the Foundations and Philosophy of Quantum Theory.* Proceedings of a Conference held at the University of Western Ontario, London, Canada, 1973. xx + 385 pp.
3. J. Bub, *The Interpretation of Quantum Mechanics.* 1974, ix + 155 pp.
4. D. Hockney, W. Harper, and B. Freed (eds.), *Contemporary Research in Philosophical Logic and Linguistic Semantics.* Proceedings of a Conference held at the University of Western Ontario, London, Canada. 1975, vii + 332 pp.
5. C. A. Hooker (ed.), *The Logico-Algebraic Approach to Quantum Mechanics.* 1975, xv + 607 pp.
6. W. L. Harper and C. A. Hooker (eds.), *Foundations of Probability Theory, Statistical Inference, and Statistical Theories of Science*, 3 Volumes. Vol. I: *Foundations and Philosophy of Epistemic Applications of Probability Theory.* 1976, xi + 308 pp. Vol. II: *Foundations and Philosophy of Statistical Inference.* 1976, xi + 455 pp. Vol. III: *Foundations and Philosophy of Statistical Theories in the Physical Sciences.* 1976, xii + 241 pp.

8. J. M. Nicholas (ed.), *Images, Perception, and Knowledge.* Papers deriving from and related to the Philosophy of Science Workshop at Ontario, Canada, May 1974. 1977, ix + 309 pp.
9. R. E. Butts and J. Hintikka (eds.), *Logic, Foundations of Mathematics, and Computability Theory.* Part One of the Proceedings of the Fifth International Congress of Logic, Methodology and Philosophy of Science, London, Ontario, Canada, 1975. 1977, x + 406 pp.
10. R. E. Butts and J. Hintikka (eds.), *Foundational Problems in the Special Sciences.* Part Two of the Proceedings of the Fifth International Congress of Logic, Methodology and Philosophy of Science, London, Ontario, Canada, 1975. 1977, x + 427 pp.
11. R. E. Butts and J. Hintikka (eds.), *Basic Problems in Methodology and Linguistics.* Part Three of the Proceedings of the Fifth International Congress of Logic, Methodology and Philosophy of Science, London, Ontario, Canada, 1975. 1977, x + 321 pp.
12. R. E. Butts and J. Hintikka (eds.), *Historical and Philosophical Dimensions of Logic, Methodology and Philosophy of Science.* Part Four of the Proceedings of the Fifth International Congress of Logic, Methodology and Philosophy of Science, London, Ontario, Canada, 1975. 1977, x + 336 pp.